HANS W. HANNAU

In the Coral Reefs of the

CARIBBEAN • BAHAMAS • FLORIDA • BERMUDA

HANS W. HANNAU

In the Coral Reefs of the

CARIBBEAN • BAHAMAS • FLORIDA • BERMUDA

IN COOPERATION WITH

ROBERT C. WORK • DENNIS M. OPRESKO • PATRICK L. COLIN • ROBERT BRODY

DOUBLEDAY & COMPANY INC.
GARDEN CITY, NEW YORK

Cover design by Bill Hays

ISBN: 0-385-05036-4

Library of Congress Catalog Card Number 74-3551

Book composed in U. S. A.
Printed in Spain
D.L.B-27900 - 1974

CONTENTS

THE PICTURES

COLOR PHOTOGRAPHS BY:

Bernd Mock: pages 26 lower (2), 29, 30, 31, 32 upper right and lower, 35, 38 upper, 39, 66 lower, 72 upper left, 74 lower right, 75, 77, 78 upper, 79, 80 lower, 98 (2), 99, 102 lower, 104 right, 105 upper, 106, 107 upper right, 109 lower, 110 upper left.

Dave Woodward: Front cover, pages 25, 34 upper, 65, 67 (2), 70 upper, 72 lower, 73 lower, 74 upper (2), 76 upper right and lower, 80 upper, 97, 100 (2), 101 (2), 102 upper, 103 (2), 108 upper left and lower right.

Armando Jenik: pages 26 upper, 27, 28 upper, 37 right upper, 40, 69, 71 (2), 73 upper right, 76 upper left, 108 lower left.

Teddy Tucker: supplied left upper and lower picture on page 36 and right lower picture on page 37.

J. Victor Neuhaus III: pages 28 lower and 70 lower.

Frederick M. Bayer: page 110 lower (2).

James W. Latourrette: pages 32 upper left and 66 upper.

Peter Vila: pages 34 lower and 107 upper left.

Ron Blakeley: pages 38 lower and 68 lower.

Robert S. Farrelly: pages 72 upper right and 109 upper.

Fran Hutchings Thorpe: page 111 lower (2).

Bronson Hartley: pages 74 lower left and 78 lower.

Mel Victor: pages 110 and 111, both upper right.

Dennis Opresko: page 107 lower.

Johannes Zachs: page 112.

Gordon Lomer: page 105 lower.

Robert Brody: page 104 left.

John Stormont: page 68 upper.

☆

INTRODUCTION

Another world of life and beauty and a new field for science have opened up for men, women and children in the last few decades. No longer is the world beneath the sea a domain reserved for intrepid adventurers. Diving has become a family pastime, and now young maidens explore coral reefs, thanks to modern diving equipment.

Snorkeling above a reef is a delight to the eyes, the skin and the spirit. Diving with SCUBA equipment is the nearest thing to flying that humans can enjoy. The mysterious sea, that has so enchanted the imagination of men down through all the ages, is opening up and revealing so much diverse beauty, and still more mysteries.

The sea itself has become the modern marine scientists' laboratory. They go down and live underwater, find the answers to questions that baffled men for millenia, clues to the origin of the earth and of life on earth. They are supported in their exploration of the blue depths by the hunger to know more about that which has been aroused in this diving generation. One of the greatest thrills for humans today is the discovery that there are still new horizons, new scenes to explore that men have never yet seen, beneath the sea.

☆

GEOGRAPHY OF THE WESTERN ATLANTIC FLOOR

The very name of the Atlantic Ocean has magic in it, for it is derived from the legendary, lost land, Atlantis. Atlantis lay beyond the Pillars of Hercules and the bath of all the western stars, and sank beneath the sea. For centuries it was marked on maps—as Atlantis, the Fortunate Islands, Antillia and Avalon. The Greeks, the Romans, the Portuguese, the Irish and the Welsh told stories of that lost land far out in the ocean.

The Atlantic has the distinction of being the saltiest of the world's oceans. The major features of the topography of the floor beneath the water are known, though much detail remains to be investigated. The ocean floor is divided by the Mid-Atlantic Range, which stretches 10,000 miles from Iceland to the Antarctic and lies about equidistance from the continents that rim the ocean. To the east and to the west of the range are great depths, with the deepest part of the ocean being about 36,000 feet.

The Atlantic trade winds blowing west along the equator create two great equatorial currents that flow from east to west. As the currents approach the Americas, the spin of the earth causes one to flow northward along the shore of North America, one southward along the South American coast. The northern current backs up in the Caribbean, flows into the Gulf of Mexico and then out again through the Florida Straits between Florida and Cuba and up along the coast as the Gulf Stream.

The Gulf Stream swings eastward toward Great Britain in the latitude of Cape Hatteras and warms the Bermudas as it flows across the Atlantic. Bermuda, summit of a submerged volcano capped by aeolian limestone, is the most northerly island group that is fringed by coral reefs in the world. The North American continent is bordered in the Atlantic by the continental shelf, the best explored part of the ocean. It is composed of sediment washed down by the rivers of the continent when the waters of the world were lower during the glacial periods. Beyond the continental shelf lies the continental slope, which descends rapidly and often steeply to depths of about 6,000 feet.

The Greater Antilles, named for the lost Antillia, are the summits of a drowned mountain chain. The Lesser Antilles, the smaller islands which separate the Caribbean Sea and the Atlantic, are of more recent volcanic origin, and earthquake activity goes on there still today. The most fascinating feature of the clear, warm waters from the Caribbean to Bermuda are the coral gardens and canyons that grow in the comparatively shallow waters that lap the shores of the islands.

THE LIVING REEF

Warm, clear water, strong sunlight and flowing currents are required by reef-building corals. All these requirements are met in the western Atlantic from Bermuda to Brazil, and it is in this wide area that some of the most magnificent coral reefs on the planet are found. The many varieties of corals are all carnivorous animals, polyps, that eat small sea creatures floating by, capturing them with stinging tentacles. Coral reefs are the skeletons of these coral polyps.

Though most people know corals as the white, exotically shaped stone that is sold in curio shops, the colors of living corals may be vivid — orange, red, green, gold. Many are brownish yellow. The soft, living, tube-shaped coral polyp forms about its surface a hard layer of calcium carbonate. This cup around the polyp is called the corallite. Colonies of living polyps form the mass of the coral reef. The colonies grow in exotic shapes that give them their popular names, such as elkhorn coral, staghorn coral, brain coral. These colonies may form enormous domes.

The coral polyps feed only at night, and during the day their stinging tentacles are pulled back into the skeleton. They eat small crabs, shrimp, crawfish and very small fish, which they shoot with microscopic poison cells. Within the reef-building corals live minute blue-green algae. The algae give off oxygen which the coral polyp breathes, and the algae absorb the carbon dioxide that the polyp gives off—a mutually beneficial relationship called symbiosis. The sex life of the coral polyp is not very sexy. Every polyp develops sperm and eggs during the breeding season. The ripe sperm leave the polyp by way of its mouth, float along and are swallowed by another polyp. The sperm fertilizes the eggs, which leave by way of the mouth when they grow as large as a pinhead. The coral larvae float away and settle down within a few days or a few weeks on hard surfaces. There they cement themselves, grow up, and become tubes with tentacles—polyps. The polyps immediately start making stony skeletons and a new coral colony is born.

Though there are small corals that grow in cold waters and at great depths, reef corals need warm water and plenty of sunlight. They grow where the water temperature is over sixty degrees and they thrive most actively in the warmest water.

The blue-green algae that live within the reef coral polyp need sunlight to survive and produce the oxygen that the polyp absorbs. Since sunlight is filtered out as it passes through water, reef corals cannot grow at depths greater than 150 feet. Reef corals are killed when silt and sand settle on them for any length of time, and they need currents and waves to bring food and oxygen. For these reasons the reef corals do not grow well in sheltered waters.

Coral reefs are described as fringing reefs, barrier reefs, bank reefs and atolls. All of these types are found in the western Atlantic and the Caribbean. Fringing reefs begin in the shallow waters close to the coastline and grow outward toward the sea because they need the wave action. The reef continues to grow outward from the land until water depths reach about ninety feet. The original inner coral reef may die in part, due to sedimentary deposits in the water inside the edge of the reef.

Barrier reefs are more complicated than the fringing reefs. They are separated from the shore by wide lagoons and channels and run parallel to the shore. They form huge masses of coral, and when the coral dies and crumbles on the seaward side it forms a suitable base for new growth. The bank reefs which are common in the western Atlantic grow on the continental shelf that was deposited when the coasts were eroded during the Ice Age. An atoll is a coral reef that has been formed on an undersea base far from land and is shaped in a rough circle which encloses a shallow lagoon. Charles Darwin theorized that atolls grow on subsiding volcanic peaks, and more than a century later evidence was obtained that proved this is true. As the reef crumbles and is eroded into sand, the sand banks grow beside the lagoon and a doughnut-shaped island is formed.

The coral reefs round Bermuda are nurtured by the warm Gulf Stream. Bermuda was once a volcano. It sank beneath the sea and was covered by limestone formed in the ocean. Coral grew on the limestone, and then was exposed when the sea level dropped in the Ice Age. This mass was eroded by wave action and formed a platform in shallow water. The present Bermuda reefs grew as the ocean warmed and the coral rose. These are bank reefs.

The eastern coast of south Florida also has bank reefs which have grown offshore on the edge of a shallow shelf that was carved from the land in the last glacial period. The reefs begin at Fowey Rock, two miles off Miami, and extend all the way to the Dry Tortugas, paralleling the Florida Keys. Cold water and sedimentation have limited the growth north of Miami, where there are only occasional patches. There is a little coral growth along the west coast of Florida, where it is also limited by cold and sediment in the Gulf of Mexico. The magnificent reefs that stretch along the Florida Keys are separated from those islands by a channel, Hawke's Channel. In the sheltered shallows between the reef and the shore there are fish-filled patches of coral.

Even more beautiful and extensive are the bank reefs of the Bahamas. They fringe islands that were atolls before the last Ice Age. Visitors to Nassau enjoy glass bottom boat trips to the coral canyons to the east and west of Nassau. The waters east of Andros, Eleuthera and Abaco have especially fine living reefs.

Every kind of living reef is found in the Caribbean. Fringing reefs surround Barbados. The Colorados barrier reefs lie off the north coast of Cuba. The longest reefs in the Caribbean are barrier reefs —one that stretches east and west of the Isle of Pines and the other that extends between Trinidad and Cape Cruz. Hogsty Reef east of Cuba is an atoll. Cay Sal Bank off the Florida Keys is a drowned atoll. Fringing reefs are found off the coast of Jamaica. In the Gulf of Mexico the principal reefs are found near Veracruz and on the Campeche Bank. Splendid bank and barrier reefs stretch for more than 125 miles along British Honduras. All of the smaller islands in the Caribbean are fringed by coral gardens.

☆

OCTOCORALLIA: Sea Fans, Sea Feathers and Sea Whips.

The coral reefs are blessed with an infinite variety of life: fish galore, sponges, molluscs, crabs, crawfish, shrimp, starfish, octopi, turtles, featherduster worms and octocorals. The fan-like, fernlike, plumose octocorals are among the most numerous and decorative creatures that adorn the seascapes of the West Indies and the western Atlantic.

These octocorals include the conspicuous and elegant lavender sea fans swaying in the swells, the sea feathers and the sea whips. All of these are gorgonians, which are among the most numerous and widely distributed octocorals that grow on the coral reefs. Gorgonians, with their horny and branching skeletons, get their name from the three snaky-haired Gorgon sisters in Greek mythology whose glance turned the beholder to stone.

The octocorals that form the fantastic sea gardens look like plants, but they are animals, colonies of polyps like the true corals. They secrete distinctive, calcarcous skeletons, as do the corals, and they too are sedentary. Some anchor themselves to rocks, reefs and shells by secreting a calcareous disc which they attach to hard surfaces. Some anchor in a muddy bottom by a fleshy stalk. Octocorals are found from the low tide line to great depths. The deepest on record was brought up from 6,250 meters.

Some are called soft corals, some leaf corals. Some are sponge-like in texture on the surface. Some are shaped like candelabra, some flagelliform. Colonies come in all sorts of shapes and colors — red, purple, lavender, amber, greenish, yellow, pink, orange. Shrimps, worms and snails live among the colonies. Fish use them as camouflage to escape their enemies. They add great color and grace to the reefs as they sway gently in the undersea currents.

Dense fields of gorgonians grow on the rocky bottom off the Florida Keys, Bermuda, Caribbean and the Bahama islands. The decorative, lacy sea fans are particularly popular among undersea collectors. Specimens may be dried, and when they are dried they are brittle. The calcareous skeletons of the octocorals contribute substantially to reef formation.

SPONGES

Sponges are also sedentary, plant-like animals and the sponges used for mopping since the time of Homer are the silk-like skeletons of colonies of parazoa. Water-bearing food is drawn through the holes in sponges. Some are soft, some are limey and brittle, and there are a variety of shapes in addition to the familiar bath sponge. The Bahama Banks were one of the world's great sources of the sponge of commerce until the 1930s, when a fungoid killed ninety per cent of the sponges in the western Atlantic. The appearance of many of the sponges found around the West Indies is suggested by their names: mermaid's glove, sea fig, brittle horny sponge, wool sponge, velvet sponge, honeycomb reef sponge, hairy yellow sponge. One of the charms of the sponges is that they are veritable sea slums. Tear a sponge apart, bit by bit, and out will come a host of tiny sea creatures — minute crabs and shrimps, brittle stars, sea worms and other wriggling bits of life.

☆

STABBERS AND STINGERS

The sea urchins, the jellyfish, the Portuguese man-of-war and the sea anemones are among the reef-dwellers that should be avoided. The sea urchin is an echinoderm with a limey skeleton shaped like a pin cushion and covered with spines. Woe to the diver who touches the black spiny sea urchin, for the long spines are needle-sharp and brittle. They stab, break off and fester. The roe of some sea urchins, obtained by breaking open the shell and scraping the pink roe from the interior ridges of the shell, is a gourmet delicacy. Sand-dollars are thin, flat members of the echinoderm family. Star fish are strong echinoderms that can pull an oyster open and eat it.

The jellyfish or medusa should also be avoided, as should the related sea anemones. Both forms are coelenterates, and both sting. On days when large patches of the sea are covered with jellyfish, go elsewhere to dive. The jellyfish is a free-swimming animal that moves with rhythmic contractions through the water. It is bell-shaped and has hanging tentacles that sting and cling to its prey. The pain of such a sting can be intense. The same can be said for the sea anemone, a sedentary relative of the jellyfish. This is a polyp with fleshy tentacles that fastens itself with a suction cup on its base to rocks. It is a beautiful thing, with its tentacles waving gently in the swells, and the uninformed swimmer or diver is tempted to pluck it. Nobody makes that mistake more than once.

The Portuguese man-of-war is a colonial jellyfish that can give a serious sting. This decorative creature has a delicate, air-filled, lavender-colored bladder that floats on the surface of the water. From it hang purple tentacles many feet in length. Groups of these deceiving creatures may float many miles over the ocean before they beach themselves on the sand and die. Observant sailors have noted that they can tack into the wind. They sting and eat small fish, but the little man-of-war fish can take refuge among the tentacles without being stung. In the same fashion, the clown fish can escape from its enemies by swimming among the tentacles of the sea anemone, where it is also unstung.

☆

MOLLUSCS

Sea shells have for ages been one of the most cherished harvests of the sea because of their beauty. Many are the inland children who first heard the sea when they put a sea shell to their ear. Beach-combing for shells is a way of life. Divers have an even better opportunity to collect rare and perfect specimens, for many are found at considerable depths. Many, such as the shell of the argonaut, or paper nautilus, are so fragile that they crumble almost as soon as they are beached. It would require a book to describe all the exotic shells found in the warm western waters of the Atlantic, and good books do exist.

Conchs are especially valued by the natives of the West Indies. Conch chowder and conch fritters are delicious, conch shells are saleable, and natives of the Florida Keys and the Bahamas call themselves conchs. These beautiful shells with their pink lining are often found between the reef and shore in areas where they have not been heavily collected. Pearls have been found in conchs.

Oysters and clams were a staple of diet of the Indians of the western world from ancient times and their camp sites are still marked by shell mounds. Varieties of these delicious molluscs are found all over the world in shallow tidal waters and near estuaries. Oysters growing on the mangrove roots along tropical shores are called coon oysters, because the raccoons love them. All are delectable, if the waters in which they grow are not polluted. Scallops are also among the tasty shellfish of the deeper waters.

Not all molluscs have shells, for squids, cuttlefish and octopuses are among the naked members of this phylum of invertebrate animals. All are edible, enjoyable if properly cooked. The squid throws up a cloud of ink when attempting to escape. The octopus changes color to camouflage itself when in flight. It can give a painful bite with its beak if grabbed, and the bite of some species is quite dangerous.

☆

TURTLES

Sea turtles are gracious about giving rides to divers and many divers have learned how to swim them to the surface. They simply grasp the front of the shell with one hand, the back with the other, and steer them to the top of the water.

Green, loggerhead, leatherback and hawksbill turtles are found in the Caribbean, the Gulf of Mexico and the Atlantic. Largest of these marine turtles is the leatherback. Most threatened is the green turtle, because green turtle soup is so eminently edible. The flesh resembles veal. Turtles lay their eggs in the sand on the beaches in April and May and return to the beaches where they were hatched to deposit their eggs. Humans find the turtle eggs with their leathery shells edible, but raccoons love them. Green turtles have become rare around Florida. Ross Witham of the Florida Department of Natural Resources has a project under way to repopulate Florida beaches. Marine turtles and their eggs are protected by Florida statutes. Green turtles are still caught commercially along the coasts of Central America and in the Caribbean, and there is a green turtle nursery on the Cayman Islands. Tortoise shell, from which the frames of eye-glasses were once made, comes from the hawksbill turtle, smallest of the sea turtles. Tortoise shell is still used commercially as an ornament and veneer, and the material comes principally from the West Indies and Brazil.

CRUSTACEANS

Succulent crustaceans, the crabs, lobsters, crawfish, prawns and shrimps, are meals prized both by man and fish. Many are found within the reefs and along the shores. They are scavengers, predators, or algae eaters and are fed on by the large fish and marine mammals. Their larvae are an important portion of the plankton which forms the diet of the giant whales. Many of the crustaceans contribute to the phosphorescence that illuminates the tropical seas at night. In tropical waters the crawfish, sometimes called Florida lobster and rock lobster, replace the Maine lobsters of cold waters. Though clawless, they are delicious, and they abound in the reefs. Shrimp, large and small and in a variety of species are also abundant. Small banded coral shrimp, which resemble elegant insects, clean the teeth and gills of big fish.

Blue crabs, stone crabs, land crabs, fiddler crabs, hermit crabs, Sally Lightfoot crabs and ghost crabs are all abundant in and around the warm ocean. Blue crabs and stone crabs are considered gourmet fare.

☆

FISHES

Many big and magnificent books have been and will be written about the fishes of the warm waters of the Atlantic and the West Indian Seas, and most divers soon acquire a good guide book that enables them to identify the colorful multitudes they see. In fact, they acquire shelves of books. The dominating characteristic of the reef fishes is that they are colorful and maneuverable, able to scoot into crevices when chased. One of the great pleasures about exploring under the sea is the discovery of the different personalities and habits of the fishes.

Groupers, for example, are stay-at-homes that live in caverns and slurp in their meals. They are one of the biggest of the reef fish. A diver off Key West was once inhaled by a jew-fish, a huge grouper. His head and air tank got stuck in the creature's throat and he fought his way out. He had little teeth marks across his chest to testify to his adventure. On the other hand, groupers can be tamed to come out and feed from a man's hand. The moray eel, a fish that looks like a sea serpent with a keel on its back, also retreats into holes and can give a mean bite to an intruder. A moray eel can also be trained to eat morsels of fish from a diver's fingers. (Better wear gloves.)

Parrot fish have fused teeth that resemble a beak. They are sand-makers, for they nibble coral and excrete it as sand. Angelfish, trigger fish and butterflyfish are among the most beautiful of the reef inhabitants. Sawfish bear their young alive. This would seem at first thought to be a do-it-yourself Caesarean, but the young are born with a plastic-like sheath around their saws. Manta rays are the waltz kings of the reefs, and their beautiful, slow swimming motion is a delight to watch. Stingrays have a barb at the base of the tail that injects a poison, and they are not fun to step on.

Sharks bear their young alive, and baby sharks huddle together as close as a pack of sardines for some months after they are born. The most important single thing for the diver to know about sharks is that they are unpredictable. All can attack, including the nurse shark that is considered harmless by divers. Millions of men and women have dived for a life-time in "shark-infested" waters without harm. One diver reported being scraped by the sandpaper-like hide of a hammerhead shark as it went past him to take a fish from the end of his spear. Some divers carry shark billies, short clubs with nails in one end, to hit inquisitive sharks on the nose. These are better than the knives with which novices equip themselves for protection. Considering how many millions of people swim and dive in the Atlantic today, it is reassuring that shark attacks are so rare.

Certainly there is a tremendous lot to be learned scientifically about fishes and the ecology of the reef, now that people can go down and live under the sea, watch the sharks swallowing the beer cans. Also, the possibility of composing vivid vignettes, profiles of the personalities of the citizens of the ocean, has barely been tapped. There is unlimited fun awaiting those who will write about "Fishes I Have Known."

☆

SCHOOLS OF FISH

Men have observed for eons that some fish are gregarious, that they tend to swim together in great schools. This schooling instinct enables fishermen to harvest big crops, ranging from small herring to large tuna. The instinct is also useful to members of the group. Schooling seems to serve as a form of protection to small fish. The large crowd of anonymous individuals confuses the aim of predators because they can't focus on an individual. Hence there is safety in numbers. Schooling also increases the chance of fishes finding others of their species that are ready, willing and able to mate.

However, fish do not school simply for protection and a better opportunity to find a mate. Sharks and barracudas go hunting in packs, as do wolves. A school of predators can herd a school of small fish into a cul de sac. Sharks demonstrate a social instinct from the time they are born, and they are born alive.

Schools also offer fish an opportunity to learn from each other. A pack of barracudas hung around under a lighthouse on a Florida reef. A spear fisherman stood on the bottom step of the lighthouse ladder and shot one. The others moved away and could not be induced to come back under the light. Marine biologists have demonstrated that goldfish learn from others of their kind. There are, therefore, numerous reasons why fish do not have to be made to go to school. Their natural social instinct is strong and beneficial.

☆

UNDERSEA NIGHT-LIFE

Night time is not a quiet time on the reefs. Some fish sleep, some do not. Some fish change color and pattern when they sleep, as though they were putting on pajamas. Sharks seem to have insomnia during the night, but then they have poor eyesight, and an excellent sensory apparatus that detects vibrations in the dark. Coral polyps and others of their ilk come out to dine at night, withdraw in the daytime. Marine scientists who have camped out at night on the reefs have made many fascinating observations never previously suspected. Among the most breath-taking delights at night, on or under the sea, is the bioluminescence, the phosphorescent streaks and sparks that light up the ocean waters.

Night can be noisy under the sea. The three-inch damsel fish chirps when he courts and pops when he fights. Fish grunt, wheeze, swish, vibrate, pop, click and groan. There is more noise on the reefs at night than there is in the daytime. Dr. Arthur Myberg of the University of Miami has made these observations in an underwater studio off Bimini. His studies are sponsored by the U.S. Navy and the National Science Foundation.

The U.S. Navy is interested in noises made in the sea. Navy sonar operators are often fooled by noises made by undersea life because they mask the sounds of submarines. Scientists of the U.S. Naval Oceanographic Office aid the Navy's antisubmarine forces by studying and identifying these sounds. The oceans have many types of noisemakers, including many types of whales, porpoises, some lobsters, snapping shrimp and countless groups of swimbladder-equipped fish. Many types of marine life, besides emitting noise, can also deflect sound waves, another aspect of bioacoustics that sends Navy biologists to sea to stop, look and listen.

Here are some of the things that the Navy has learned about noisy sea creatures:

The snapping shrimp, very small—smaller than a man's little finger—can be found throughout the world's oceans, principally over coral, sponge, shell and rock bottoms. The species was first "heard" by World War II submarines, who reported the noise to be similar to "sizzling, like burning brush or frying bacon." Marine biologists, on investigating, found that the noise was created by literally thousands of snapping shrimp. An individual shrimp, they reported, makes a sound similar to a cork being pulled from a bottle, a noise that is probably produced when the shrimp withdraws a prominent lobe or plunger into its single pincer-like claw.

The drum fish, say the Navy's experts, is a carnivore that lives in warm, sandy nearshore areas throughout the world. One of the best known of sonic fishes, its drumming noise was first referenced by marine biologists in 1908. The drumming, scientists noted after extensive study of the fish, is created via the swimbladder, primarily in the male. Some types of drum fish also whistle through their teeth. The drum fish's sonic activity usually increases during breeding season, when congregations of males sing in loud nocturnal "choruses." Twenty-two types of drum fish have been identified in American waters off the Atlantic coast.

The croaker is a very close relative of the drum fish. With both species included in the same family, their habits and body structures are quite similar. The croaker's noise, however, resembles a bull frog's croak more than the sound of a drum, and the female Atlantic croaker, unlike most of her cousins, often sings with the males during mating season.

The toad fish, a carnivorous coastal inhabitant of warm seas, is a versatile noise maker. It often generates two sounds—a coarse grunt or hoarse, raucous growl, as well as a clear musical note or "bloop" that sounds much like the blast of a boat whistle or foghorn. The growl, according to marine biologists' studies of the fish, may indicate a warning to other members of its species, annoyance, fear or aggravation. The whistle, on the other hand, is probably a male sound, emitted during breeding. The toad fish is a bottom fish of sluggish habits, preferring to dwell among rocks, rubbish and weeds close to shore.

Singled out as a noisemaker by early writers, sea robins were often called "crooners," "pipes" or "grumblers" and figured as the subjects in some of the first studies in marine bioacoustics. Like many of the drum fish, croakers and toad fish, the armoured sea robins sing via their swimbladders. Some emit loud barks as they compete for territory or food, which many of the bottom-dwelling fish dig out of the mud with their fins. Sea robins can be found in both shallow and deep waters.

The smiling porpoise, or bottle-nose dolphin, is one of the most intelligent of all marine creatures. This small cetacean is called a porpoise along the Atlantic coast to distinguish it from the dolphin fish, no relation. An air-breathing mammal, the porpoise lives in schools, principally in the Atlantic and the Mediterranean, where it feeds on small fish and shellfish. Observed since ancient times, it became a prominent figure in bioacoustics during the 1950s. The porpoise, while swimming near the surface, apparently "talks" to other porpoises through a combination of squeaks and whistles made through its blowhole. In addition to its surface "chatter", it seems to use sound, a series of clicks, to chart a course through ocean depths.

The porpoise, according to biologists' studies, can transmit the clicks at frequencies up to 170,000 cycles per second, a frequency that is ten times higher than man can hear. It emits the clicks in a scanning motion and then appears to listen for echoes, all the while swimming forward, and seems to know instinctively that the time interval between transmittal and reception denotes the distance to the bottom or to a submerged object. While Navy biologists are fascinated with the porpoise's surface "chatter," they are even more interested in its natural sonar capabilities and hope to use knowledge gained from on-going studies of the porpoise in rescue and salvage operations.

☆

PARTNERSHIPS BENEATH THE SEA (SYMBIOSIS)

The mutual aid pacts that have developed between different species are fascinating aspects of life on the coral reefs. A diver may be lucky enough to see many different examples. A little neon goby, a fish with a brilliant blue stripe running the length of its body, will swim in and out of the mouth of a moray eel without coming to harm. Why has this peaceful coexistence come to pass?

The little goby is the principal cleaner fish in the Caribbean and Western Atlantic, and has many other species of fish as well as moray eels among its customers. A goby sets up a cleaning station, and customers wait in line. A customer fish gives a signal that it wants to be cleaned by assuming a position with its mouth wide open. The little cleaner goby swims up and into the mouth, carefully removing parasites from the gills and mouth. When the customer decides that the cleaning is adequate, it closes its gill cover once or twice and the cleaner swims out and away to another waiting customer. The goby makes a meal of the parasites that grow on gills and skins, the customers are relieved of the itching they cause. Banded coral shrimp and Peterson shrimp do the same thing, and are among the cleaners of the Caribbean and Western Atlantic.

This sort of mutual aid pact is called symbiosis, which the dictionary defines as: "The intimate living together of two dissimilar organisms in a mutually beneficial relationship." Another definition might well be: "You scratch my back and I'll scratch yours." Participants in a symbiotic relationship may get transportation, food, protection and cleaning.

The remora is a hitch-hiking symbiont that rides around on sharks and big whip rays. This fish has a dorsal fin that has developed into a suction cup that enables it to fasten itself to the hide of a shark. In addition to the ride and the protection offered by the shark, the remora shares in the shark's meals, gobbling up bits that fall from the shark's prey. What's in it for the shark? The remora cleans while it rides, eats skin parasites on the body of its host. Tiny parasitic crustaceans that have been found on the skin of sharks have been found in the stomachs of remoras.

The sea anemone has developed two pacts with very dissimilar creatures. The sea anemone is an animal with fleshy, beautifully colored tentacles. It resembles a plant, and is usually found attached to coral or rock. Its tentacles can inflict a painful sting, and it eats little fish that it stings when they touch the gently waving tentacles. Sea anemones have been observed riding around on the backs of hermit crabs. A crab will loosen the fibers that attach the sea anemone to a rock and place it on its back. The sea anemone camouflages and protects the hermit crab. In return, the anemone, which would otherwise be stationary, has a better chance of getting food when it rides around on top of the crab.

There is a small clownfish, a species of Amphiprion, that can retreat for protection among the tentacles of the sea anemone and live there without being stung. The anemone eats food dropped by the clownfish. The clownfish will attack and bite anything that approaches the anemone, including divers. Other fish, such as butterfly fish, eat anemones, but the attacking clownfish will drive them off. Peterson shrimp also live unharmed among the tentacles of the sea anemone. Thus both symbionts protect each other.

Another crab, called the decorator or sponge crab, camouflages itself very deftly with other living species. This crab takes pieces of sponge and algae and cultures a growth on its shell. The cultured sponges are held in place on top of the shell by the fifth pair of legs. The camouflaging growth gains mobility in this relationship.

There are endless examples of what is called commensalism among different sea creatures. In this relationship one creature gains protection or food from the relationship, does not injure its host, but apparently does nothing for the host. The Portuguese man-of-war is a siphonophore. This creature has a large, air-filled, lavender-colored sac which floats on the surface of the ocean. Mariners report that they have observed the creatures tacking into the wind. From the sac hang long tentacles which can inflict a most painful sting. The Portuguese man-of-war dines on little fish that it stings. There is, however, the little man-of-war fish, a species of the genus Nomeus, that can swim among the tentacles and be protected without being stung. What, if anything, the man-of-war fish may do for the Portuguese man-of-war is not known.

An intriguing example of communal living, "let's everybody room together next semester," can be found within shells in which hermit crabs live. The hermit crabs live in the shells of sea molluscs that have died and left them vacant. There is also a very tiny crab, called a spotted or porcelain crab, that may be found living in the shell carried around by the hermit crab. There can be as many as five of the minute porcelain crabs sharing the shell. Barnacles may attach themselves to the outside of the shell. It is indeed a condominium-like situation.

☆

COLORS BENEATH THE SEA

The rich colors of the creatures that swim among the coral reefs are a source of endless delight to divers. Marine observers who have camped out at night under water have noted that some fish put on pajamas, change color at night. In addition to pleasing the eye of the beholder, colors beneath the sea serve various other purposes. Colors may be used by fish as identification, camouflage, signals and means of establishing territoriality and attracting mates.

Bright colors are found among fish that have a home range, a sense of territoriality. Such fish tend to be aggressive toward members of their own species, to chase them away from their territory. It is an advantage to a species as a whole to have a wider range, more food. Thus the bright colors that say to a member of the same species, "No trespassing!" result in a wider distribution of that species.

Vivid colors not only serve to drive a fish's own kind away, they draw potential mates together. Brilliant colors and patterns are frequently found where there are many different species in the same families, such as butterflyfish. They are badges of a particular species, recognition signals. Since cross-breeding generally is infertile or produces defective offspring, and distinctive color patterns minimize crossbreeding, this function of color is useful.

There is a great generation gap in the sea. Many parents eat their children, minimizing competition. Thus young fish of many species are often very different in coloration and pattern from adults. The small fish dress to blend themselves in with their surroundings, and two generations can live together in harmony.

Cuttlefish and octopi use color as camouflage, and can change to blend into the background more skillfully than chameleons. A fascinating camouflage pattern has been developed in the four-eyed butterflyfish. Many blennies attack the eyes of other fish. The four-eyed butterflyfish has a spot on each side of its tail that looks like an eye. Blennies have been observed attacking the false eyes of the butterflyfish. Camouflage by color is useful not only for protection but also for the predator that is sneaking up on its prey. Trumpet fish, for example, can adopt a sand-colored camouflage that blends in perfectly with the sea bottom when they are stalking prey.

☆

WORLD OF ADVENTURE

Throughout the Caribbean there are marvelous coral gardens, some in crystal clear water so shallow that non-swimmers wearing face masks can float above them on inner tubes and enjoy the fish-filled world. Deeper waters filled with fish, wrecks, treasure and beauty invite experienced adventurers.

Buck Island Reef National Monument in the U.S. Virgin Islands has an underwater trail through which divers can swim in thirty minutes. Uninhabited Buck Island lies off the northeast coast of St. Croix. A few hundred yards offshore is a handsome barrier reef of elkhorn coral. Many small boats operate between St. Croix and Buck Island. Spearfishing, the possession of spearguns and collecting marine life is prohibited by regulations of the National Park Service. There are picnic facilities and restrooms on Buck Island, but no camping is allowed. There is a delightful nature trail and fine beaches on the island. The Virgin Island National Park northeast of St. John is also a realm of scenic sea-scapes.

Other attractive reefs in the American Virgins are those at Thatch Cay, Little St. James, Long Point, Congo Cay and Outer Brass. Anegada, in the British Virgins, has offshore reefs that are littered with sunken ships. One of the nice things about ship wrecks is that fish like them and they are productive hunting grounds for spear fishermen.

Buccoo Reef off Tobago is one of the world's finest submarine gardens, and the shallow waters allow novices to float and observe the beautiful world with face mask and snorkel. Sand Cay Reef in the bay of Port-au-Prince in Haiti is another wonderful reef in comparatively shallow water. Magnificent reefs, plus 225 known wrecks of sunken ships, are found off the Cayman Islands. Puerto Rico has splendid reefs offshore. Prickly Pear reef off Antigua is popular with skin-divers. Jamaica, Trinidad, Curaçao and many other islands all have fine fringing reefs. Many of the smaller islands, recent volcanos, rise quite steeply from great depths, but most of them have coral patches in the waters on the leeward side of the islands.

Bermuda is blessed with underwater beauty in its fringing reefs, and sunken wrecks offshore date back some centuries. They have yielded treasure and legends.

The Bahamas, 700 green islands stretching in a 750-mile arc through 100,000 square miles of the clearest water anywhere, are a paradise for divers. The island chain begins fifty-five miles off the coast of Florida and trails through the Atlantic down to the northern edge of the Caribbean. The islands have numerous excellent diving resorts. A mecca is the Underwater Explorers Society in Freeport, with a training pool, library, technical laboratory, sauna and game room. There also is the Cousteau Museum.

Andros is fringed by the third largest barrier reef in the world. It stretches for 120 miles. At the edge of the Tongue of the Ocean it is twelve feet deep on the inside, 6,000 feet on the outside. The Blue Holes of Andros are openings that plunge more than 200 feet down into the reef, and are seen from the surface as deep blue against the sunlit shallows. Strong currents running counter to the tide set them apart from any similar formations elsewhere. The Wall, the Andros Dropoff, plunges 100 feet down. Small Hope Bay wreck, in seventy feet of water, is home for many groupers.

Off Abaco there is the Hole in the Wall. Spanish Cay offers some of the most beautiful diving waters in the Bahamas. Miles of sea garden in Pelican Cay National Park are protected by the Bahamian Government, and no spearfishing or coral collecting is permitted. Around the ocean-side reefs of Abaco there are numerous wrecks.

Off Rum Cay lies the wreck of the "Ocean Conqueror," a gunboat with 101 guns, dating to 1861. Elkhorn coral fringes many recent large wrecks around Mayaguana. The Berry Islands boast fine reefs and good diving. Around Stella Maris on Long Island are limpid waters. Especially attractive are the Cape Santa Maria ocean heads in about fifty feet of water and the Clarence Town blue hole. Crooked Island's Hogsty Reef is the only true atoll in the Bahamas.

The Exuma National Land and Sea Park includes a golden sea garden that stretches for twenty-two magnificent miles. Underwater attractions of the Exumas also include the mystery cave on Stocking Island, fish-filled and unexplored beyond 500 feet; the Thunderball Grotto at Staniel Cay, the Highborne Cay wreck of a 1560 privateer, Wax Cay cut, Bond's Cay cut, the Exuma Sound Dropoff and the Stocking Island blue hole.

Off Grand Bahama on Treasure Reef the $1,200,000 Lucayan treasure was found in 1964. Divers also enjoy the Wall, Black Forest ledge about a mile off the south shore of the island, the Zoo Hole and the caves on the south shore.

Eleuthera offers an exhilarating underwater ride through Current Cut. Boiling Hole off Current, Egg Island reef, the Six Shilling channel, Powell's Point, Yellow Banks and numerous unusual wrecks are favorite Eleuthera underwater attractions.

Wrecks abound off the Biminis, and they include a one-time gambling ship. Turtle Rocks, Little Isaacs, East and West Brothers, Gingerbread Grounds, El Dorado Shoals and Mosell Bank are among the areas around the Biminis that fascinate divers. Here, it is claimed, man-made structures have been sighted beneath the sea.

Great diving waters round New Providence include Lyford Cay dropoff, North Rock blow hole, the Rose Island reefs, Goulding Cay reefs, Gambier deep reef, Green Cay heads, Booby Rock channel, south side minireef and the Clifton Pier dropoff. Blizzards of fish and beautiful coral abound.

☆

TREASURE HUNTING

Gold has glinted in the dreams of men and treasure hunting has filled their fantasies for countless centuries. Since gold-hunting Spaniards first came to the New World, the backdrop for many of the fantasies has been the green islands and the blue waters of the West Indies. Bermuda and Florida, too, have become encrusted with legends of treasure.

The trouble about sunken treasure is that it becomes encrusted with coral as well as legends. Coral larvae need a firm footing on which to grow, and they begin to cover wrecks not long after they go down. It does not take many decades for a ship to become completely camouflaged with stone-hard coral growth and deeply covered with sand.

Shakespeare, who certainly never saw the reefs round Bermuda, described nicely what happens on them when in 1611 he wrote "The Tempest." Shakespeare had been reading Sir George Somers' account of a ship wreck on the Bermudas in 1609. Shakespeare had Ariel sing:

"Full fathom five thy father lies;
Of his bones are coral made;
Those are pearls that were his eyes:
Nothing of him that doth fade
But does suffer a sea-change
Into something rich and strange."

The diving gear available since World War II immediately gave new impetus to treasure hunting in the western Atlantic and the Caribbean. There have been plenty of finds to keep the dreams alive. Fact, not fiction, has led troops of treasure hunters to the "Spanish Sea," the Caribbean, the Bahamas, and to the waters round Florida and Bermuda.

Cortez found gold in the Aztec Empire in Mexico, but nowhere near as much as Pizarro found in the land of the Incas, now Bolivia, Peru and Ecuador. Much silver and precious gems were also looted from the Incas. This was the stuff of which Columbus had dreamed, this was the metal of the Potosí mines and of El Dorado, the legendary city of gold. Early in the sixteenth century this treasure began to flow across the Atlantic to Spain.

Each year a Spanish fleet, the Plate Fleet, assembled in the New World to carry treasure to Spain. Spanish galleons and merchant ships annually gathered at Porto Bello and at Veracruz to be loaded with gold, silver, pearls and precious stones brought across the Isthmus on muleback in leather sacks. They sailed for Havana, hub of New Spain. Due to prevailing winds they sailed north from Central America to the coast of Louisiana, and then curved south along the west coast of Florida and thence to Havana.

After all the ships of the fleet had gathered in Havana harbor, they set forth for Spain together. There is safety in numbers, and they knew that English, French and Dutch pirates, privateers and buccaneers haunted the seaways ready to plunder them. There might be almost a hundred ships in the fleets that sailed for home. They took the same route each year, past the Florida Keys, north up the Gulf Stream between Florida and the Bahamas, along the North American coast until, with the Gulf Stream, they swung east in the latitude of Bermuda. Thence the prevailing westerlies sped them to the Azores and then to Spain.

All along the shores and reefs that border this route sunken treasure may be found. Coral reefs are graveyards as well as sea-gardens. The Spaniards persisted in sailing from Havana during the hurricane season. The galleons were clumsy, not maneuverable, and many ships sank in storms. Others simply disappeared. Modern treasure hunters not only count on luck, rumors and metal detectors, they also search through old archives and maps in Spain, which may tell them the approximate locality in which the Plate Fleet of New Spain sank in a certain year, or where the ships were when a hurricane hit in another.

Some treasure hunters have been delighted to find in wrecks not only gold, silver and jewels of the New World, but also precious porcelain and other works of art from the Orient. The Spaniards held the Philippines by the middle of the fifteenth century, and traded with China, Japan and the Indies. A big Spanish galleon, the Manila Galleon, brought the Oriental wares across the Pacific to the Isthmus, and they were carried by mules to the eastern shore of the Isthmus.

One of the nicest things about undersea treasure hunting is that the scenery is so pretty the divers don't feel cheated if they find nothing. Many ships of all descriptions have gone down on the reefs of the Caribbean, Bahamas, Florida and Bermuda. The reefs off the Florida Keys are so treacherous that eight great lighthouses of the world were built along them in the nineteenth century. Luck still today plays an important part in finding sunken treasure troves. But some of the lucky men have known where to look and what to look for, and they haven't looked at the end of the rainbow. They have looked on reefs, they have looked for ballast stones, for straight lines that might be coral encrusted cannons, for old anchors. And some have searched for years before they found their prize.

Consider Teddy Tucker of Bermuda, one of the lucky ones. He began diving in helmet and diving dress in 1938. For years he worked as a commercial diver on the wreck-strewn reefs round Bermuda. Not until 1955 did he find his greatest treasure, a gold and emerald cross which he retrieved from the Spanish ship "San Pedro," that went down in 1594 on Bermuda's north reefs. The "San Pedro" yielded $180,000 in gold, jewelry and artifacts, and one single gold ingot was valued at $31,000. The "San Antonio," a Spanish galleon that sank in 1621 on the west reefs of the island, also contained gold, jewelry and other treasures. There are many wrecks still unexplored around Bermuda.

Luck went over the side of the boat with Gary Simmons in 1965 when he dived into twenty feet of clear water about a thousand yards offshore from the Lucayan Beach Hotel on Grand Bahama. He had spotted what he thought was a coral encrusted anchor, but it turned out to be an anchor encrusted with silver coins, mostly crude pieces-of-eight struck from flat bars of silver. Simmons and three friends, who shared with him in owning and operating the water skiing and skin diving school at the hotel, brought up a million dollars' worth of treasure in the form of silver coins minted about 350 years ago and bearing the insignia of Philip the Fourth of Spain. Three more treasure sites are reported to lie just offshore of Grand Bahama Island.

Successful treasure hunting is not new in the Bahamas. Captain John Smith, he who took the first settlers to Virginia in 1607, headed an expedition that salvaged twenty-six tons of gold bullion from a wrecked Spanish galleon in the clear waters off Turks Island in the seventeenth century. Captain William Phips brought up tons of Spanish treasure between Inagua and Hispaniola in 1687. Legend says that the Haitian tyrant Henri Christophe hid a great cache of gold on the northern end of Inagua island, where he is said to have built a summer palace. Inagua is nearer Haiti than any other Bahama island. A twenty thousand dollar Spanish silver ingot was found off South Abaco Island in 1950.

There is a dangerous reef seven miles to the west of the Caribbean end of Crooked Island Passage, the seaway for large ships heading between the Bahamas and the West Indies for South America. It is surely a happy hunting ground for treasure seekers, for here there is an untold number of ancient wrecks. This dangerous reef west of tiny Castle Island in the Bahamas was named by the Spaniards "Mira Por Vos," "Look Out for Yourself."

Treasure hunters must be licensed and wrecks registered, according to the law of the Bahamas. 'Twas not always thus. Those islands were a refuge of English pirates for almost two centuries. It was said of the Bahamas Governor Cadwallader Jones in the seventeenth century that he "highly caressed those Pirates that came to Providence." Nassau, on New Providence Island, the capital of the Bahamas, was also the headquarters of numerous notorious pirates — Edward Teach, alias "Blackbeard," Calico Jack Rackhan, Steve Bonner and the lady pirates, Mary Read and Anne Bonny. A buried pirate treasure is said to have been hidden on Fortune Island in the Long Cay group in the southern Bahamas.

Luck steered Kip Wagner to a Spanish coin uncovered by a storm on a Florida beach near Sebastian Inlet in 1964. But competence and sophisticated equipment led Wagner, who formed Real Eight, Inc., and Mel Fisher of Treasure Salvors, Inc., to find the offshore wreckage of the Plate Fleet of 1715. From these ships they salvaged three-and-a-half million dollars in treasure, one of the great troves of all time. Some of this find is very attractively displayed in the Museum of Sunken Treasure at Canaveral, near the coast where it was found. The museum is a delight and an education to potential treasure hunters of all ages.

Florida, because it parallels the route of the Plate Fleet and lies in the path of hurricanes, is a mecca for treasure hunters. Treasure hunters of Miami in the 1930s developed a device that could detect metal, which did not work well above water. They sold it to the U.S. Government and it became the mine detector of World War II. Much more successful devices have been developed since then, and the aqua-lung has spurred the search. The State of Florida requires that treasure hunters have permits to salvage their finds in State waters, and the State takes twenty-five per cent of the treasure trove. Excavating wrecks has become part business, part science, but it still takes luck.

All the sunken treasure that is found does not glitter. In 1968 a Spanish galleon, the "San Jose," part of a fleet sunk by a hurricane in 1733, was found off the Florida Keys by divers Tom Gurr and Rudolph Palladino. The wreckage was buried in the sand in thirty-five feet of water beyond the three-mile Florida underwater boundary. It was part of a twenty-one ship fleet that included three galleons, and it left Havana on July 13, 1733. Of these ships, seventeen, including the galleons, went down along the Florida Keys. Spanish records show that the Spaniards brought salvage parties from Cuba within two weeks after the ships sank, but that they recovered less than half of the valuables from the wrecks.

Diver Art McKee in 1948 found the first wreck of a galleon and opened a treasure museum on Plantation Key with his finds. In 1958 Olan Frick found a second galleon, "El Infante." In their search that resulted in discovering the "San Jose" in 1968 Gurr and Palladino used a copy of an old Spanish map and a couple of electronic devices to pinpoint the wreck. Their months of salvaging with a team of divers cost them more than $100,000, the treasure hunters said. The Smithsonian Institution called the treasure that was found "an extremely valuable collection."

Marine archaeologists of the National Park Service in 1972 used a proton magnetometer, a sophisticated metal-detecting device, to locate twenty-four shipwrecks on the coral reefs in the Dry Tortugas. One of these is believed to be the "Nuestra Señora del Rosario," one of several ships that went down when a hurricane hit the Plate Fleet that sailed from Havana on September 4, 1622. The Spaniards, according to archives in Spain, located the site twenty days after the wreck and salvaged 500,000 pesos of silver.

The wreck lies partly buried in sand and encrusted in coral in water that is six to eighteen feet deep. The Dry Tortugas are the site of Fort Jefferson National Monument and are sixty-eight miles west of Key West. George R. Fischer led the National Park Service team of diving archaeologists. The Service hired a Washington firm, Earth Satellite Corp., to fly a small plane with highly specialized cameras over five square miles of the waters around the Dry Tortugas. The film showed a coral reef shaped exactly like a ship one hundred feet long. When it was plotted on the map it fell right on the site of the wreck that had previously been discovered by the divers.

On July 4, 1973, a team of divers discovered, in twenty feet of water in the Marquesas Islands forty miles southwest of Key West, a sunken treasure for which men had been searching for 350 years. The silver ingot found by the team, working for Treasure Salvors, is from Nuestra Señora de Atocha, one of eight Spanish ships sunk by a hurricane in the Straits of Florida in 1622.

The divers have also recovered other silver bars, silver pieces of eight, numerous guns and cannon balls, religious jewelry and pottery and copper ingots from Cuban copper mines. Their successful search continues. The silver ingot has been authenticated as coming from the Atocha, which carried about forty tons of treasure and other goods from the Americas.

A state salvage investigator, Dennis English of the Division of History and Archives, who was monitoring the search to insure the State of Florida gets its twenty-five percent, said that he believes the discovery could be "one of the most spectacular," historically and monetarily, in Florida.

The waters round the Caribbean islands are rich in sunken wrecks. This sea was the cockpit of warring European powers in the sixteenth and seventeenth centuries. The Spaniards salvaged some of the cargo of ships that went down, using Lucayan Indians, great divers. By the seventeenth century, there were many secret maps of sunken treasure drawn from the accounts of survivors.

Jacques-Yves Cousteau knew exactly what he was looking for when in 1968 he explored the Silver Bank, a coral reef northeast of the island of Hispaniola. The Silver Bank has been a graveyard of ships since the time of Columbus. Here the "Nuestra Senora de la Concepcion" sank on November 3, 1641, after being crippled by a hurricane. It was from this ship that Sir William Phips in 1686 took millions in silver plate, gold coins, silver ingots, pearls, emeralds, rubies, diamonds, golden statues and crystal cups. Cousteau was looking for the wreck of this galleon, which had previously been spotted by a treasure-hunting friend.

Cousteau and his team of experienced divers worked for months on a wreck they thought was the treasure-filled Spanish galleon. The first indication of the wreck they spotted was a strange-looking coral form that resembled a cannon. A cannon it proved to be, and nearby were several more cannons and large anchors. The divers hacked away at coral in thirty-five feet of water, sending up four or five tons of coral a day through a vertical suction tube powered by an air compressor. They did find a medal and cross of gold, but after months discovered they were working on an eighteenth century vessel. Any profit from the expedition was made from the book "Diving for Sunken Treasure," written by Cousteau and Philippe Diolé, one of the great books about Treasure Diving.

☆

continued on page 43

Descriptions of the following pictures

page 25

DIVING INTO A CORAL GROTTO

This photograph shows the typical brilliant purplish blue of the deep ocean framed by the walls of a coral grotto with rich growths of sponges, gorgonians and corals.

page 26 (upper picture)

GROUP OF TARPON BETWEEN SMALL SILVERSIDES

Above the reef, a school of tarpon (Megalops atlantica) *swims through an aquatic Milky Way of small silversides* (family Atherinidae). *Such views of overflow of life are unforgettable diving memories.*

page 26 (lower pictures)

SHALLOW REEF: CORAL AND SEA FANS

A photographer taking a picture of a brilliant blue sea fan. Many reefs, like the one in the picture, are so shallow that the sunlight plays on the coral and the beautiful blue sea fan in the foreground.

ELKHORN CORAL, BLUE TANG AND SERGEANT MAJORS

A beautiful blue tang (Acanthurus coeruleus) *and sergeant majors* (Abudefduf saxatilis) *swim through a forest of elkhorn coral* (Acropora palmata). *Staghorn coral* (Acropora cervicornis) *is visible in the background. The blue tang belongs to the family of surgeonfishes, all of which possess one or more sharp spines on either side just in front of the caudal fin.*

page 27

BLACKBAR SOLDIERFISH

Many sponges splash this rocky ledge with warm shades of red and orange. Adding their own reds are four blackbar soldierfish (Myripristis jacobus) *and one squirrelfish* (Holocentrus rufus) *in the center. The latter may appear to be running with the wrong crowd, but both species belong to the family Holocentridae. Both species occur in Florida and the West Indies, the squirrelfish occurring also in Bermuda.*

page 28 (upper picture)

CHASING A GREAT BARRACUDA

A diver approaches a great barracuda (Sphyraena barracuda) *above a gorgonian covered reef. The branching forms seen here represent several species of plexaurid gorgonians; the sea fans are* Gorgonia ventalina. *The great barracuda is found in all tropical seas with the exception of the eastern Pacific.*

page 28 (lower picture)

SHARK BETWEEN STAGHORN CORAL

That is the unmistakable outline of a shark (family Carcharhinidae), *swimming over the sand between dense growths of staghorn coral.*

page 29

WHITE SEA URCHIN

A strikingly beautiful sea urchin (Diadema antillarum). *The white spines stand out in strong contrast against the black shell or test of the urchin. Spines of this species are normally black, and a few short black spines may be seen if one examines the photograph carefully. The dull red sponge in the background is* Haliclona rubens. *Although* Diadema *occurs throughout the West Indian region, it is not found on the west coast of Florida. However, it does occur on the Flower Garden Banks off the Texas coast.*

page 30

A REEF COMMUNITY: BANDED BUTTERFLYFISH, GORGONIANS, SPONGES AND STAR CORAL

Banded butterflyfish (Chaetodon striatus) *swim in a garden of corals, gorgonians and sponges. The towering coral is the large star coral* (Montastrea cavernosa). *The numerous polyps covering the branching tube sponge* (Callyspongia vaginalis) *are not part of the sponge but are zoanthids* (Parazoanthus *sp.*). *It is a typical picture of a reef community.*

continued on page 41

DIVING INTO A CORAL GROTTO

GROUP OF TARPON BETWEEN SMALL SILVERSIDES

SHALLOW REEF: CORAL AND SEA FANS

ELKHORN CORAL, BLUETANG AND SERGEANT MAJORS

BLACKBAR SOLDIERFISH →

CHASING A GREAT BARRACUDA

SHARK BETWEEN STAGHORN CORAL

WHITE SEA URCHIN

A REEF COMMUNITY: BANDED BUTTERFLYFISH, STAR CORAL, TUBE SPONGES AND GORGONIANS

REEF LIFE: FRENCH GRUNTS BETWEEN STAGHORN—AND ELKHORN CORAL

OCTOPUS AND SCORPIONFISH

MOON JELLYFISH

MATING LOGGERHEAD TURTLES

CONEY AMONG SEA FEATHERS AND SEA FANS →

BASKET STARFISH ON SEA FEATHERS

LIMA CLAM

RED SPONGE WITH BRITTLESTARS AND BLUE WRASSE

TEDDY TUCKER SCANNING THE WATERS FROM A BALLOON FOR TREASURES BENEATH THE SURFACE

EMERALD CROSS, RECOVERED BY TEDDY TUCKER FROM SPANISH SHIP SAN PEDRO, LOST 1594 ON NORTH REEFS OF BERMUDA

TIMBERS OF SPANISH GALLEON SAN ANTONIO, WRECKED 1621 ON WEST REEFS OF BERMUDA, DISCOVERED BY TEDDY TUCKER

TREASURES BENEATH THE SEA

DIVER INSPECTING THE WRECK OF THE RHONE, VIRGIN ISLANDS

CORAL GROWTH ON ANCHOR WINCH OF FRENCH FRIGATE L'HERMINIE, WRECKED 1838 ON N.W. REEFS OF BERMUDA, DISCOVERED BY TEDDY TUCKER

MILLION DOLLAR SPANISH TREASURE FOUND 1965 NEAR LUCAYA, GRAND BAHAMA

YELLOW DAMSELFISH BETWEEN SPONGES AND STAR CORAL

CARDINALFISH WELL PROTECTED AMONG THE SHARP SPINES OF A SEA URCHIN →

A TRIO OF SPOTTED MORAYS

DIVER IN FRENCHMAN'S GAP,
VIRGIN ISLANDS

Descriptions of the foregoing pictures

continued from page 24

page 31

REEF LIFE: FRENCH GRUNTS BETWEEN STAGHORN AND ELKHORN CORAL

A school of French grunts (Haemulon flavolineatum) *seeks shelter between colonies of staghorn coral* (Acropora cervicornis) *in the foreground and elkhorn coral* (Acropora palmata) *just behind. The French grunt is one of the most common fishes on the reefs throughout the West Indian region. Staghorn and elkhorn corals require clear water with good circulation, although, of the two, staghorn is more tolerant of less favorable conditions. Hence in some areas, especially the Florida Keys, staghorn coral grows somewhat closer inshore than does elkhorn coral.*

page 32 (upper left)

OCTOPUS AND SCORPIONFISH

An octopus shows his underside. A well-camouflaged spotted scorpionfish (Scorpaena plumieri) *lies on the bottom. Octopuses are difficult to identify in the field, but the two most common West Indian reef species appear to be* Octopus vulgaris *and* Octopus briareus. *The beautiful, spotted* Octopus macropus *may be found in grassy areas near reefs, but it is uncommon to rare in most localities, the grass flats of Bermuda being a notable exception. The scorpionfishes are sluggish and often difficult to detect among the bottom rubble. Their venomous spines are capable of inflicting intense, long lasting pain, and sometimes serious infection ensues.*

page 32 (upper right)

MOON JELLYFISH

An animal of ethereal beauty, the moon jellyfish (Aurelia aurita) *appears to be made of gossamer. Although the sting of this jellyfish is not intense, great numbers of them drifting over the reefs can make diving uncomfortable for especially sensitive persons.*

page 32 (lower)

MATING LOGGERHEAD TURTLES

Loggerhead turtles (Caretta caretta) *on a connubial bed of sand and turtle grass* (Thalassia testudinum). *Later the female will lay as many as 130-150 eggs at a time in nests dug high on sandy beaches.*

page 33

CONEY AMONG SEA FEATHERS AND SEA FANS

Beautiful color and forms of the reef: A coney (Cephalopholis fulva) *seeks shelter among corals, sea feathers* (Pseudopterogorgia bipinnata) *and sea fans* (Gorgonia ventalina). *The coney is one of the most common groupers in Bermuda, Florida, and the West Indies.*

page 34 (upper)

BASKET STARFISH ON SEA FEATHERS

Basket starfish (Astrophyton muricatum) *perch on the branches of sea feathers* (Pseudopterogorgia acerosa) *where they spread their elaborate arms to entrap plankton and minute bits of organic matter carried by the currents. This photograph was taken at night. During daylight hours the basket starfish curls into such a tightly wound mass of arms that it is scarcely noticeable on the sea feather.*

page 34 (lower)

LIVE LIMA CLAM

How beautiful and interesting is such a living mollusk, a white-tentacled lima clam or file shell (Lima scabra *form* tenera). *Although file shells can swim for short distances, they spend most of their time attached by byssal threads to rocks and corals. This species is common in southeast Florida, the West Indies, and parts of the Gulf of Mexico.*

page 35

RED SPONGE WITH BRITTLESTARS AND YOUNG BLUEHEAD WRASSE

A red sponge (Mycale *sp.*) *is home for several brittle stars* (Ophiothrix lineata) *only the arms of which may be seen extended in the center of the photograph and also protruding from the cloacal opening of the sponge. The fish above the sponge is an immature bluehead wrasse* (Thalassoma bifasciatum).

pages 36-37

TREASURES BENEATH THE SEA

Descriptions with pictures except page 37 (upper).

DIVER INSPECTING THE WRECK OF THE RHONE IN THE VIRGIN ISLANDS

A diver explores the skeleton of a sunken ship, which is completely overgrown with many species of sponges and a few gorgonians. It is the wreck of the British mail ship "Rhone", which sank in a storm in 1867 off Salt Island east of St. John, Virgin Islands. She is 360 ft. long and lies in 35 to 80 ft. of water.

page 38 (upper)

YELLOW DAMSELFISH BETWEEN BRANCHING AND GREEN SPONGES AND STAR CORAL

A young yellow damselfish (Eupomacentrus planifrons) *swims before a branching sponge* (Iotrochota birotulata). *Just below the small, lavender-tinged colony of star coral* (Montastrea cavernosa) *is an amorphous specimen of a fragile green sponge* (Haliclona viridis). *This sponge may be mound-shaped, lobate or branching, but the green color is distinctive.*

page 38 (lower)

A TRIO OF SPOTTED MORAYS

With mouths agape, a trio of spotted morays (Gymnothorax moringa) *presents a formidable barrier at the entrance to their lair. Morays can be found from Bermuda to the Bahamas, Florida to the Caribbean and Brazil.*

page 39

CARDINALFISH WELL PROTECTED AMONG THE SHARP SPINES OF A SEA URCHIN

Symbiosis between fish and sea urchin: Tiny belted cardinalfish (Apogon townsendi) *seek protection among the dangerously sharp spines of a black sea urchin* (Diadema antillarum), *well hidden between coral and sponges.*

page 40

DIVER IN FRENCHMAN'S GAP, VIRGIN ISLANDS

A diver explores a submarine canyon. All the beauty of submarine colors and forms is brought out in this excellent photograph. The conspicuous gorgonians are Iciligorgia schrammi. *The small, pink, coral-like growths are coralline algae.*

☆

EXPLORING THE LAST FRONTIER

Pictures by International Underwater Explorers Society, photographed by Dave Woodward

Venus, love's lady, was born of the sea, said the Greeks, and they have been exploring the birthplace of love and life for three millenia. In Homer's time Greek divers were diving for oysters and pearls, searching for barriers around harbors, recovering treasures. They used a snorkel-like device in those days, a tube with one end in the diver's mouth and one end above water attached to a floating bladder. Alexander the Great is said to have sat on a throne in a barrel of glass under the Aegean Sea, observing sea monsters and such. Greeks have been great sponge divers throughout recorded history, and became adept in the nineteenth century in the use of diving helmet outfits to harvest sponges. Greeks today are exploring Homeric cities that sank beneath the Mediterranean in earthquakes long ago.

The best diver who goes down with only his own lungful of air can go no deeper than about 140 feet, and stay under no longer than about two minutes. Men have been making diving bells and diving suits for centuries to enable them to stay longer and see more. The diving bell, it is said, was invented by Roger Bacon in 1250. Edmond Halley, who gave his name to Halley's comet, invented a diving bell in 1691 that enabled him to stay down at sixty feet for an hour and a half in the bell. In 1830 Augustus Siebe invented a diver's helmet and dress connected by a tube to a hand-cranked pump that supplied compressed air, and similar helmet-and-hose outfits have been used ever since then by divers for salvage, for gathering sponges and pearls, and for undersea work. A self-contained underwater breathing apparatus (SCUBA) that used a cylinder of oxygen and air was invented in the nineteenth century.

In that same century divers who worked too long and too deep learned about the bends, the painful and sometimes crippling and fatal disease that attacks those who breathe air under pressure too long. Its symptoms are extreme pains in muscles and joints, choking and itching. It sometimes results in paralysis, sometimes in death. A Frenchman, Paul Bert, investigated. He found that when a diver's body is under pressure nitrogen in the air he breathes goes into solution in his blood stream. As long as the diver remains under pressure he has no problem. If he comes up from too far too fast the nitrogen in his blood will bubble, clog his arteries and hurt, cripple or kill him. Bert found that if the diver comes up slowly he can eliminate the nitrogen from his blood stream. The development and testing of decompression tables has followed.

In the 1930s United States swimmers diving for sport began to use homemade goggles, fashioned after the goggles that the skilled divers of the Orient had developed. Face masks were imported from Japan. Spear fishing equipment was modeled after the Hawaiian slings of the Pacific, and a new sport was born. Rubber fins to increase speed and range were developed by the French Navy in the 1930s. Rubber frogman suits were developed in World War II to protect divers from the chill.

The underwater world really opened up for millions of people with the invention of the aqua-lung in 1943 by Jacques-Yves Cousteau and Emile Gagnan. However, few people heard of the aqua-lung until Cousteau's book "The Silent World," appeared in 1953. Then everybody who could read and swim yearned to try it. The virtues of the aqua-lung are that it affords greater mobility and safety than other diving devices, and that it is affordable by people of moderate means. A boy can make enough money mowing lawns to rent an aqua-lung and a tank of air, and if he saves his money he doesn't have to mow too many lawns before he can buy a regulator. A girl can do as well baby-sitting.

The diver using the aqua-lung carries compressed air stored in one, two or three tanks strapped to his back. The tanks are attached by a tube to a regulator and a mouthpiece. Secret of the effectiveness of the regulator is that air flows to the diver only when he inhales through his mouthpiece. It is safer than the previous self-contained diving

NEMO (U.S. Navy)
Acrylic sphere with almost 360° viewing capable of carrying two men to 2000 ft. while being towed. An umbilical to the surface supplies power.

devices, which used oxygen and air, because oxygen has certain dangers. A person doesn't have to be a very good swimmer to enjoy diving with an aqua-lung.

The trek to the sea began. It has not been lemming-like, though some divers have not returned. Aqua-lung diving is not entirely without dangers. It is still necessary to decompress on ascent from considerable depths after long dives, and some skilled divers have been crippled with the bends because they could not or did not avoid ascending too fast. The limit of the dive with this apparatus is about 250 feet, for down there in the twilight blue of evening lurks the deadly nitrogen narcosis, what Cousteau calls "the rapture of the depth." It is a sort of elegant and lethal drunkenness that makes the diver want to go on down, down, down, or to take his mouthpiece from his mouth and give a fish a breath of air. It is this deadly rapture that limits aqua-lung diving to about 250 feet deep.

A diver who is going down in rather shallow water for a comparatively brief tour in the coral canyons will need only a face mask, an aqua-lung, flippers and a weight belt. Had he not a weight belt he would have to fight to go down, and he adjusts the lead weights on his belt until he is weightless underwater. If he plans to go to any depths for any considerable length of time he will also need a depth gauge, a decompression table, a compass and a rubber wet-suit. Even in warm August waters on the reefs a human can become chilled if he is undersea for any considerable length of time.

The yearning to go deeper has been felt by so many SCUBA divers, and this yearning has spurred the extraordinary development of other diving devices in the last two decades. In the American Revolution a one-man underwater boat was built, but the first practical submarine was designed by John Philip Holland, an American, in 1900. The first extensive "practical" use for which men built submarines was warfare. Navy submarines were initially designed as shallow divers for warfare against surface vessels, and did not allow people to look around and about under the sea.

Vehicles designed specifically to observe beneath the oceans were developed more than fifty years ago. William Beebe dived half a mile down in 1934 in a bathysphere, a steel ball lowered by a cable. Jacques Piccard is one of the pioneers in deep diving. Bathyscaphs can go almost seven miles down, but are not maneuverable like the deep diving submarines. These submarines, called D/RV's (deep research vehicles) were developed for oceanographic research in the 1960s. They go much deeper than Cousteau's "sea fleas," which dive to two thousand feet. Many of the deep diving submarines can be piloted by anyone who is sufficiently competent to drive a car.

Ed Link and Cousteau pioneered in developing underwater habitats, in which men and women can live for weeks at a time. Divers move in and out of these undersea stations freely, work on the sea floor. Their body tissues become saturated with the gasses they breathe, and they can stay down almost indefinitely, though they require days of decompression when they return to the surface. Jon Lindberg, whose father is the pioneer of the air who first flew across the Atlantic alone, has been a pioneer of undersea exploration in the Bahamas.

The Underwater Explorers Society makes its international headquarters on Grand Bahama Island. It has marine biology and color photography laboratories, a research library, an undersea museum, and a luxurious sauna and gym. There are also expert instructors and training pools for novices. In the Bahamas' Tongue of the Ocean, with its great depth, the U.S. Navy Atlantic Undersea Testing and Evaluation Center (AUTEC) tests and tracks undersea devices. Here also J. Louis Reynolds, aluminum executive, conducts underwater research. Exploration of the Tongue of the Ocean has revealed that there is a shelf of solid sea fossils down there that is twenty thousand feet thick. The coral barrier reef off Andros Town is exceeded in size only by the Great Barrier Reef of Australia.

Grand Bahama Island was the site of a futuristic oceanographic experiment in 1970. The Perry Foundation placed Hydro Lab more than a mile off shore. This undersea habitat enabled researchers to live underwater while conducting studies in biology, geology, ecology and medicine. The Hydro Lab was designed to be much less expensive than more heralded undersea experimental habitats, and as a consequence during its undersea life Hydro Lab, since it first went into operation in 1966, has logged more time as an underwater habitat than any other system. One of the pioneering activities was the first direct dry underwater transfer of divers from a submarine to an underwater habitat. This operation established the feasibility of using submarines as taxis from the surface to deep sea laboratories.

Another futuristic experiment was the first use of an underwater fuel cell. This powercell attached directly to Hydro Lab provided complete air, water and power requirements. This enabled Hydro Lab to exist without the cumbersome and expensive support ships that are necessary with traditional habitats.

Undersea research is being carried out by many universities and governmental agencies throughout the western Atlantic and the Caribbean. One of the great oceanographic complexes of the world is situated on Virginia Key in Miami. Cooperating in research there are the University of Miami Rosenstiel School of Marine and Atmospheric Science, the Tropical Atlantic Biological Laboratory of the National Marine Fisheries Service, the Atlantic Oceanographic and Meteorological Laboratory of the National Oceanic and Atmospheric Administration and the Miami Seaquarium.

Perry Hydro Lab *(see special story on page 129)*

On Bermuda there are the Bermuda Biological Station for Research, Inc., at St. George's West, the Government Aquarium at the Flatts, and the Geophysical Field Station of Columbia University on St. David's Island. In the Bahamas, in addition to the governmental agencies and private research projects previously mentioned, marine studies are being conducted by the Lerner Marine Laboratory on Bimini and by the Ministry of Agriculture and Fisheries. Barbados Department of Science and Research and the Bellairs-McGill Research Institute, both at St. James, engage in oceanographic research. In Jamaica the University of the West Indies has marine laboratories at Kingston and Discovery Bay. The Caribbean Marine Biological Institute of the Netherlands Antilles is stationed on Curaçao. The University of Puerto Rico's Institute of Marine Biology is stationed at Mayaguez.

What do the undersea explorers do, what do they report? The scientific observations that have been made are of inestimable value. In addition, there are so many details and discoveries that intrigue the imagination.

Perry Cubmarine "*Shark Hunter*"

One deep-diving submarine was attacked by a sword fish. Starfish, sea pens, worms live on the deep bottom of the sea. At depths of two thousand feet there are shrimp ten inches long. A thousand square miles of rocks rich in manganese have been examined on the Blake Plateau, a half mile down off the coast of Georgia and Florida. What looked like an oceanful of squid was seen about a thousand feet down. Porpoises, which must breathe air, have been observed diving to depths of six hundred feet. (How do they decompress so successfully?) Red coral of gem quality has been found off Hawaii. Diamonds washed to sea have been found in the waters off southwest Africa. Huge beds of calico scallops were located 150 feet under the surface off Cape Canaveral. The deepwater submarines Aluminaut and Alvin recovered an H-bomb that fell in an air crash and sank 2,550 feet in the ocean.

Divers have discovered layers of squids, lantern fish and shrimp-like creatures. Under the Gulf Stream black sands that contain titanium, gold, chromium and other elements have been found. Microtectites—glassy particles that indicate a heavenly body collided with the earth almost a million years ago—have been retrieved from the sea bottom. Scientists camping out under the ocean have discovered hot spots on the sea floor, where water is ten times saltier than the sea. The research vessel Alvin sank and was brought up from 1,540 meters of water ten months later. The bologna in the sandwiches was pink in the center though gray on the outside. Though the apples were wrinkled, they showed no decay. The bouillon in the thermos bottle tasted fine. The deep sea is better than a refrigerator in preserving things, it seems.

☆

Perry-Link Deep Diver. *The Taxi to the Deep: 22 feet long with look-out chamber on top, and two pressure compartments, front for pilot and observer, back for two divers, who can get in and out of the "Deep Diver" at their will.*

SPONGES OF THE REEFS

Robert C. Work
Research Scientist, School of Marine and Atmospheric Science, University of Miami

While the corals and gorgonians of the Caribbean reefs offer their subdued earth tones and pastel shades, the sponges are largely responsible for the vivid colors of the reef structure.

On the more shallow portions of the reef, the brilliant yellows, reds, and oranges of the encrusting sponges flare from the rocky outcroppings of the reef framework and from the bases of the living corals. In addition to the showy encrusting species, there are some brightly colored species of larger size growing throughout the shallow reef, but many of the larger sponges are more interesting in form than in coloration. There are solitary tubes, multiple tubes, vases, and hemispheres, as well as upright or sprawling branching forms. Many are essentially amorphous, jutting suddenly from the reef in no fixed concept of design.

On slightly deeper portions of the reef, other species of sponges begin to appear, often in greater numbers and larger sizes than those seen on the reef top. Here the sponges make a greater contribution to the overall architecture and color of the reef. On the deepest parts of the reef slope, sponges abound, with enormous hollow cylinders and gigantic vases often dominating the entire scene. Many beautifully colored forms are found here, although this may not be apparent to the diver who is diving without an artificial light source.

One cannot expect the sponge communities of every reef in the Caribbean region to be in complete accord with the above picture, for local environmental factors may alter the species composition or numbers of individual sponges considerably. An abundance of sponges may be found on a shallow patch reef, while other reefs nearby may harbor only a few conspicuous sponges. Deeper portions of the reef slope in some localities may have relatively few, if any, of the giant sponges so typical of other deep areas. Nevertheless, whatever reef or depth the diver chooses, he can be certain that sponges of some kind are present.

Sponges belong to the phylum Porifera. They are riddled with pores, canals, and chambers through which water currents flow. The incurrent pores are always small and abundant, while the excurrent openings, known as oscules, are relatively few in number and much larger, often being very conspicuous. Most sponges possess an internal skeleton of spicules made from silicic acid or calcium carbonate, or spongin fibers made from a protein secretion; or the skeleton may be made from both spicules and spongin fibers. The spicules fall into two major categories, megascleres and microscleres. The former are the larger spicules, although these are so small in most sponges that details can be seen only with a microscope. Microscleres are much smaller, some requiring very high power for viewing. Spicules vary in shape from simple needle-like structures to elaborately branched objects of great beauty, and in some species the microscleres appear as hundreds of sparkling stars when viewed through the microscope. Sponges are without a mouth, digestive system, nervous system, or special organs. Life functions are carried out by cells which act essentially independently of each other. In spite of this, water currents are somehow maintained in established directions, and some sponges display definite but usually slow contractile ability. However, these contractions are not due to a cooperation of cells triggered by nerve impulses but to many individual cells responding simultaneouly to the same external stimulus.

Until quite recently it was believed that there were only two basic types of sponge cells, the flagellated choanocytes or collar cells, which maintain the water flow through the sponge, and the free-wandering amoebocytes or amoeboid cells. The latter, however, were thought to be differentiated for special functions and were thus given a number of secondary names to identify them in the occupation of these special roles. Present-day sponge cytologists, while still recognizing choanocytes and amoebocytes, have determined that many of the cells formerly considered to be modified amoebocytes are, in fact, distinctive special cell types; but some differences of opinion are held among these specialists as to just what roles certain of these cells play. At any rate, the digestion of food, skeletal secretion, and other functions are carried out by these specialized cells. Certain of these cells give rise to sperm and eggs. The eggs are fertilized internally and develop into flagellated larvae inside the sponge, from where they escape through the oscules and settle elsewhere to become small sponges. Some sponges also reproduce asexually by means of the formation of little membrane covered aggregations of cells called gemmules.

The complicated details of basic internal architecture and the many different types of spicules and fibers shall not be approached here in great detail. However, these features are very important to the sponge taxonomist. Even with such characters as the internal architecture, spiculation and fibers, in addition to the external form and color, sponges in general are very difficult to identify. Although there are a number of common, conspicuous species that may be recognized readily, there are far more that are difficult to determine because of their nondescript or variable external appearance (or perhaps close resemblance to other species) combined with spicules that are indistinctive or even absent, as in the keratose sponges. Environmental conditions may greatly alter the external appearance considered typical of many species. To further complicate the picture, many sponges incorporate into their skeletons quantities of foreign debris, sand grains, and spicules of other sponges. In addition some of the descriptions of sponges in the earlier literature are very vague, and many different names may be found for the same species.

Sponges by their very nature have less tangible attributes than most other animals, and for this reason one finds unusual, though often helpful, analogies used by authors to describe them. They have been compared to cheese, whole wheat bread, and many other edible and inedible substances; but perhaps the most puzzling is the statement by one author that the sponge "looks and feels a bit like lukewarm snow."

There are four main classes of sponges. These are the Calcispongiae, the Hyalospongiae, the Demospongiae, and the Sclerospongiae. The Hyalospongiae, the glass sponges, have elaborate skeletons of silica spicules, which are all six-rayed or are modifications of the six-rayed spicule. Glass sponges live in deep water and will not be encountered on the reef. The Calcispongiae possess spicules of calcium carbonate, usually in the form of calcite. These spicules may have one axis, or they may be three- or four-rayed. A few small species of Calcispongiae do occur on the Caribbean reefs.

The Sclerospongiae comprise a most unusual class of sponges. They have skeletons of siliceous spicules, organic fibers, and calcium carbonate in the form of crystalline aragonite. The dense deposition of aragonite in these sponges renders them unexpectedly heavy and gives them a coral-like appearance. Six species of Sclerospongiae have recently been discovered on the rugged deeper portions of Jamaican fore-reef slopes, where they are reported to be abundant enough in some places to be a significant contributor of structurally consolidated calcium carbonate to the base of the reef.

Of greatest interest to the diver are the sponges of the class Demospongiae, which are by far the most abundant and conspicuous sponges of the shallow waters of the West Indian region. These are sponges with either siliceous spicules (though never six-rayed) or with spongin fibers, or with both. Those with only spongin fibers are called keratose or horny sponges, and it is among these that the important commercial species are found. Some species of keratose sponges will pick up and include in their skeletons the spicules lost by nonkeratose species, however. The class Demospongiae also includes some sponges which have no proper skeleton at all.

Many of the West Indian sponges range north to southeast Florida, the Bahamas, Bermuda, and the offshore banks of the Carolinas. Many range into the Gulf of Mexico, where great numbers are found on the reefs in the northeast Gulf of Mexico between Tampa and Apalachicola, Florida. Incredibly, some species endure very cold winter temperatures in very shallow water at St. Theresa near Apalachicola. Reefs off Laguna Beach just west of Panama City, Florida, seem to be the last outposts for a number of Caribbean species in Florida waters. West Indian species are also found in the western Gulf of Mexico on the Flower Garden Banks and other banks off the Texas coast and on the reefs of southern Mexico. The reefs off the west coast of Florida are quite unlike those of the Florida Keys, the Bahamas, and the Caribbean. In addition to the many Caribbean sponges found on these reefs, the corals, gorgonians, and other groups are represented by a number of West Indian species; but there is also an admixture of a decidedly less tropical element, and on many reefs there are only low limestone outcroppings with more sponges and gorgonians than corals. The corals are usually not large, and ivory coral, *Oculina,* dominates. Nevertheless, these reefs, sometimes termed simply "sponge bars" by the Greek divers, are beautiful and well worth exploring. The greatest problem is their inaccessibility.

While exploring a lagoonal patch reef, a diver is quite likely to observe large, brownish, vase-shaped sponges growing around the periphery of the reef. Along with these may be rather large, black, cake-shaped sponges and lobed or branched, brownish sponges with darker oscules. These three species belong to the genus *Ircinia.* They are covered with numerous, small, pointed projections called conules, but these are much more coarse on the cake-shaped species than on the others. They are tough but resilient to the touch, and all share a very unpleasant, fetid odor. These sponges are *Ircinia campana,* the common vase sponge; *Ircinia strobilina,* the cake sponge, and *Ircinia fasciculata,* the lobed stinker sponge. It is interesting to note that the name *Ircinia* stems from the Latin word hircinus, from which also stems the English word hircine, meaning "smelling of goats." In some areas *Ircinia strobilina* is called the loggerhead sponge, but this name is more properly applied to *Spheciospongia vesparia,* which is also cake-shaped, black, and grows to an enormous size, even in shallow water. It lacks resiliency, however; and although it may have a roughened surface, it lacks true conules. It may also be recognized by its sieve-like arrangement of oscules in the center of the upper surface. Although one should not attempt to detect the odor of *Ircinia* under water, these sponges, as well as all other sponges, should not be removed from the bottom unless needed for research purposes. Sponges are an important part of the reef habitat; and with few exceptions, when dried, they are unattractive and often odiferous souvenirs.

Spheciospongia and the species of *Ircinia,* as well as many other sponges found on or near the inshore patch reefs, are also found in lagoonal areas quite apart from the reefs proper and even far into bays where the salinity is sufficiently high and water circulation is good. In the Bahamas, the Florida Keys, Central America, and Caribbean Islands the commercially valuable species of sponges are not found on the coral reefs but on relatively quiet banks or in lagoons and bays. Commercial sponges are not found on Caribbean Islands that do not have extensive areas such as these. Yet the same commercial species are found in the Gulf of Mexico quite far off the Florida coast and at considerable depths. Among the commercial species, the most important are the wool sponge, *Hippiospongia lachne;* the yellow sponge, *Spongia barbara;* and several varieties of the grass sponge, *Spongia graminea.* These are keratose sponges with very resilient fibers, numerous conules, and usually brown to black exteriors. The intensity of the external pigment of many sponges, especially those of the order Keratosa, is greatly affected by the amount of light to which the growing sponge has been exposed. Sponges of the genus *Ircinia* are sometimes mistaken for commercial sponges; but this genus, though keratose, has debris filled fibers which lose their elasticity upon drying.

There are many other conspicuous sponges of the lagoonal areas and inshore patch reefs. The maroon-red, branching or massive *Haliclona rubens* may be abundant on some reefs. Several other species of *Haliclona* are clothed in lavenders and purples, while *Haliclona viridis* has a distinctive shade of light but bright green. Equally distinctive is the beautiful sky-blue coloration of specimens of *Dysidea etheria,* which are readily distinguished from the deep blue encrustations of *Tricheurypon viride.* Tall, slender, sparsely branched specimens of *Verongia longissima,* sometimes mistaken for gorgonians by divers, grow in great numbers in some areas. This species is not hollow like some of the species of *Verongia* from deeper water.

The notorious fire sponge, *Tedania ignis,* may be very rare or absent on well developed coral reefs, but it is often very common inshore. It varies from bright yellow-orange to red and often has oscules that are elevated like little volcanoes. The fire sponge is chemically highly irritating and should be carefully avoided. If one is not certain of its identity, any orange or red sponge should be left alone.

A particularly interesting sponge of the shallows is *Geodia gibberosa*. This sponge possesses a hard cortex and has its oscules grouped together in the center of the upper surface. Although its own color is a dirty white to medium grey, this is rarely observed; for specimens are almost invariably overgrown with other sponges, usually *Haliclona viridis*. The hard surface of *Geodia* is due to densely packed, round, tuberculate microscleres called sterrasters.

Ubiquitous in lagoons and reef areas is the peculiar chicken-liver sponge, *Chondrilla nucula,* which grows as small, globular masses in grassy areas but may form thick, rather extensive encrustations in areas of rock and coral. This sponge is mottled on its upper surface in tans and browns, while the basal parts are usually light grey. As the name implies, it looks and feels something like chicken-liver but perhaps a bit firmer. It almost totally lacks a skeleton, possessing only scattered microscleres.

Several species of *Spirastrella* form small, hard, red encrustations on the reefs; and *Placospongia* may encrust fairly extensive areas with its peculiar pattern of cracks demarking the polygonal areas of its surface. *Placospongia* is red-brown in color and stone-like in consistency. Many of the brightly colored encrusting sponges are very difficult to identify. Some of them are keratose, and many of the orange and red forms belong to the families Microcionidae, Ophlitaspongiidae, and Axinellidae. By no means, however, are all members of these families exclusively encrusting forms.

Very destructive to the reef are the boring sponges of the genus *Cliona,* which excavate extensive canals through limestone rocks, shells, and corals. Some species of *Cliona* are rather drab in color, but others impart a brilliant orange or red coloration to the host object. Occasional reports of large red corals may be traced to advanced permeation of *Cliona* in boulders or dead coral heads.

Although the common tube sponge, *Callyspongia vaginalis,* occurs in very shallow water, the species begins to appear in abundance at slightly greater depths. This species is highly variable, ranging from solid, terete forms to solitary and multiple tubes or vases. The surface may appear quite smooth or greatly roughened by conules and spine-like projections; and the color ranges from grey-green to drab blue-green, sometimes with tints of lavender. Another tube or vase-shaped sponge, *Callyspongia plicifera,* shuns the most shallow depths. It is found in mixtures of delicate shades of pink, lavender, purple, blue, and orange. It is quickly recognized by its iridescent quality and the deep convolutions and plications on the outer surface. *Dasychalina cyathina,* the fringed vase sponge, is cylindrical to vase-shaped with a distinctively fringed rim and hues of blue, green, or purple. Tube sponges of the genus *Verongia* are readily distinguishable from the aforementioned three species of sponges. Species of *Verongia* are much more fleshy in structure and grow to enormous sizes on the deeper portions of the reef. *Verongia fistularis* is a smooth to roughened, hollow cylinder of bright yellow with umber or greenish tints. *Verongia lacunosa* is similar, but the yellow is more muted by other colors, and the outer surface of the sponge is conspicuously convoluted and pitted.

In the genus *Mycale* there are several species that are sub-tubular, often more like hollow domes. Colors in these forms range from dull maroon to fuchsia and cerise. *Mycale laevis* is quite different. At depths of about five fathoms and deeper, this sponge thickly encrusts the entire lower surfaces of flattened colonies of living reef corals.

Very common on many reefs is *Iotrochota birotulata,* which usually assumes an erect or sprawling, bush-like form with free or coalesced branches that are often terminally lobate. This sponge is purple-black with a definite green sheen on the branches. *Ianthella ardis* also has a bright green sheen, but the underlying color is yellowish to reddish-brown.

There are a number of shrub-like or tree-like, branching sponges on the reef. The dull to bright red *Microciona juniperina* is highly variable in shape but often resembles a small, shrubby juniper tree. *Higginsia strigilata* is also shrub-like but grows larger than *Microciona juniperina,* is usually more orange-red, and has more ridges and convolutions on the rather fan-shaped ends of the branches. *Axinella polycapella* and *Homaxinella rudis* are red, tall growing species with elongated cylindrical branches. The former has a velvety smooth surface, while the surface of the latter is quite roughened.

On the deeper reef, in addition to the earlier mentioned large species of *Verongia,* there are a number of other large growing species. *Xestospongia muta* always attracts attention. It usually occurs as a very large, shallow-bottomed vase or a hollow-barrel whose sides are quite rugose and even buttressed in older specimens. Its color ranges from light ocher through shades of lavender and pink. Although this sponge is quite firm, portions may be broken away with very little effort. The genus *Cribrochalina* is also represented on deeper portions of the reefs by large, irregular vases, but these are darker in color than *Xestospongia* and may even be purple-brown. Another large vase or basket sponge of the deeper reef is a species of *Geodia,* which appears to be distinctive from the shallow water *Geodia gibberosa.* The color varies from a very light to a medium brown, and it is exceedingly hard in consistency.

Fibulia massa is an encrusting to massive, easily crumbled sponge, resembling *Xestospongia* in texture to some extent, but very rarely vase-shaped. The color is a mixture of dull yellow and red-brown shades. This sponge produces a chemical irritant even more severe than *Tedania,* the fire sponge; and it should definitely always be avoided. Unfortunately, it is not always easily recognizable. While *Fibulia* and *Tedania* are chemically highly irritating, many other sponges have sharp, outwardly directed spicules which may penetrate the hands of a careless diver. Particularly good examples are sponges of the genus *Cinachyra,* which look like small, bristly, yellow-orange cakes with numerous round perforations on the surface.

There are several conspicuous species of *Agelas* on the deeper reef. The reddish-brown *Agelas sparsus* is massive, often strongly lobed or bulbous, and has meandering grooves on its surface. *Agelas conifera* may form large masses with hollow protuberances, the orifices of which are yellow. A yellow coloration is also found on portions of the outer surface, especially near the base; elsewhere the surface is purple or brown. Other species of *Agelas* occur in brilliant shades of orange. The distinctive spiculation of *Agelas* consists solely of handsome, stylus-shaped megascleres that have whorls of short spines surrounding the length of the shaft.

Careful searching of the reefs, at various depths, may reveal small, bristly, dirty-white, ampule-shaped sponges. These are *Scypha ciliata,* one of the calcareous sponges. Another calcareous sponge sometimes found on the reefs is *Leucosolenia canariensis,* which forms beautiful, lemon-yellow, latticed masses up to several inches in diameter.

The sponges mentioned herein make up only a small percentage of the species which may be encountered on the Caribbean reefs. Many of the species, even some rather conspicuous ones, are undoubtedly undescribed species, which have yet to be given their scientific names. Due to the inherent difficulties in working with this group and the small number of scientists engaged in sponge systematics, many species are likely to remain unnamed for some time.

In addition to their contribution of beauty, sponges play a very important role in reef ecology. The destructive role of boring species has already been mentioned. Many species offer temporary refuge for many animals and permanent shelter for many more. The surfaces of many species are often covered with minute entoprocts or small anemone-like zoantharians of the genus *Parazoanthus.* Large sponges may harbor extraordinary numbers of invertebrates. Over sixteen thousand snapping shrimp of the genus *Synalpheus* have been reported from a single large specimen of *Spheciospongia vesparia,* the loggerhead sponge. Sponges are also home for flatworms, nemertean worms, nudibranchs, brittle stars, and often great numbers of polychaetes. The latter are particularly well represented in sponges by the families Eunicidae and Syllidae. A beautiful, glassy-spined brittle star, *Ophiothrix lineata,* appears to be confined to sponges, being found especially frequently in *Callyspongia, Dasychalina,* and *Mycale.* Sessile snails of the genus *Vermicularia* are often found in sponges, and the sponge oyster, *Ostrea permollis,* lives embedded in the surface of *Stelletta grubii* and perhaps a few other species. In addition to giving temporary shelter to transient fishes, some of the territorial species may establish residence in or around sponges; and some gobies seem to have definite associations with certain sponges.

Sponge crabs, *Dromidia* and related genera, tear off a living piece of sponge; and with their hindmost pair of legs they carry it closely held to and completely covering the carapace. These crabs sometimes use compound ascidians, especially *Eudistoma hepaticum,* in the same manner. Biologists in the field often mistake *Eudistoma* and other compound ascidians for sponges, and one specialist was once on the verge of describing *Eudistoma hepaticum* as a new species of sponge. It is little wonder that *Dromidia* so often makes *Eudistoma* an alternate choice. Other crabs, especially certain of the spider crabs, attach fragments of living sponges, as well as other growths to their carapace and legs. These effectively camouflaged animals are often referred to as decorator crabs.

Probably more different animals utilize sponges for food than we are presently aware. It is known that a number of fishes do include sponges in their diet, and some of the angelfishes, though eating a wide range of foods, seem to have a preference for sponges. Sponges are also eaten by some starfishes, sea urchins, and mollusks. Very soft-textured sponges are the food for certain nudibranchs, and are a part time diet for some of the cowries, *Cypraea.* Top shells, *Calliostoma,* have also been observed eating delicate sponges.

Sponges are becoming increasingly important in pharmaceutical research. Certain species have produced antibiotic compounds, and others are showing some promise in cancer research. Among the West Indian sponges that have aroused the interest of medical researchers are *Cryptotethya crypta, Chondrilla nucula Haliclona viridis, Haliclona subtriangularis, Ircinia strobilina,* and various species of *Agelas.*

If a diver should make an unusual observation involving a sponge, and he feels that it is important to have the sponge properly identified, it is not necessary to remove the entire sponge from the reef. He should take careful note of its shape and color, then cut off just a small portion of the sponge. The cut should go somewhat deeper than just the surface, for spiculation varies from the exterior to the interior in some species. If alcohol is not available for preservation, the specimen should be frozen or dried. Formalin should never be used for preserving sponges. Even when collecting museum specimens, it is recommended that the sponge be carefully cut, rather than torn, from the reef. In this way one may leave a base, from which the sponge will regrow. A conservative approach by both the layman and diving scientist will be greatly appreciated by the inhabitants of the reef.

☆

CORALS AND CORAL REEFS

DENNIS M. OPRESKO
Museum of Comparative Zoology, Harvard University

Photographs © F. G. Walton Smith

To a skindiver a coral reef is a fascinating mosaic of dazzling colors, strange forms and shapes and constant motion, but to a marine biologist a coral reef represents one of the most complex and delicately balanced communities in the sea. Built gradually by literally millions of minute coral animals encased in stone, many reefs rest on coral foundations sometimes thousands of feet thick and thousands of years old.

The solid substrate formed by the corals alloys for a tremendous diversity of marine life, and the entire reef community is directly or indirectly dependent on the corals or the coral rock, for shelter, for food or simply for a place to attach and grow. Marine algae grow in, around, and on the corals and some even live within the tissues of the coral polyps. Sessile animals like sea fans and sea whips (gorgonians), sea anemones, sponges, bryozoans (moss animals), plume worms, tunicates (sea squirts) and oysters, to name just a few, attach themselves to the solid coral rocks; while crabs and lobsters, starfish, sea worms and sea urchins, snails and octopus, plus a multitude of tropical fish hide under ledges, and in the holes, cracks and crevices of the reef. There are some organisms such as sponges, clams, worms and even barnacles which are specially adapted to tunnel through the coral limestone. When weakened by such organisms, huge mounds of coral may be toppled over by the force of the sea. At the same time, however, the reef is constantly being rebuilt and repaired, by the corals themselves and also by encrusting bryozoans, tunicates, hydrocorals and calcareous algae. In addition, chemical processes occur whereby the coral debris is cemented back together to once again form a solid hard substance.

The reef is not the only area of sea bottom influenced by coral growth. Coral debris, from the size of boulders to fine sand, may be carried seaward by currents, tides and wave action to form the sediments and talus slope of deeper water, or it may be transported landward to add to the rocky platform of the reef flat. Finer calcareous debris settles in pockets or depressions on the sea bottom or is washed inshore to form sandy beaches.

On the seaward slope of the reef, on the reef flat and on the sandy bottom around the reef, one finds communities of marine organisms different from those on the reef itself. In areas of soft sediment marine grasses and algae often carpet the bottom. These plants support still different biological communities. In addition, ancient reef platforms, which are no longer in areas suitable for optimum reef growth may serve as a substrate for communities of attached organisms such as gorgonians, sponges and algae. These communities are usually found in relatively shallow water where there is little sediment build-up. They often cover extensive areas of the sea bottom, sometimes quite far from the more typical reef areas.

Thus, reef corals provide the basic substrate for a diverse number of marine communities. In turn, the marine plants and animals which occupy these different habitats, may add considerably to the reef sediments by way of their calcareous skeletons, such as shells of mollusks and forams, the spicules of sponges and gorgonians and the skeletal plates of echinoderms and calcareous algae. Furthermore, these plants and animals often serve as a source of food for various reef inhabitants. Many fish and invertebrates leave the safety of the reef at night to prowl the surrounding sand and grass flats for prey. Sessile reef animals, such as corals and anemones, on the other hand, often feed on the planktonic young of the sand and grass dwellers. In addition, the juvenile stages of many reef animals are found away from the reef in other types of habitats. In this way the reef and the surrounding communities form a complex and intricate marine ecosystem, and this ecosystem has evolved largely as a result of the coral animal's tremendous ability to extract dissolved salts from the ocean and convert them into massive limestone reefs.

THE CORALS

For a long time it was generally believed that corals became hard as stone only when they were removed from the sea. In fact some people think that the word coral originally meant "that which hardens in the hand." The idea that corals harden on exposure to air was disproved by an enterprising Frenchman of the 18th century who had the novel idea of diving into the Mediterranean and examining living corals firsthand. It was also in the 18th century that it was finally accepted as fact that corals were actually the work of tiny marine animals. Previous to that time, corals were thought to be either minerals, or plants, or strange combinations of plant and animal.

The coral animals which are responsible for forming the vast reef systems of the world belong to the phylum of animals called the Coelenterata. The phylum is divided into three classes; the Hydrozoa, the Schyphozoa, and the Anthozoa. The first group contains the hydroids, hydrocorals *(Millepora* and *Stylaster)* and jellyfish-like organisms such as the Portuguese Man-of-War. The second group contains the true jellyfish. Reef corals are classified in the Anthozoa along with other coral-like animals including the sea anemones and sea fans. There are two subclasses of anthozoans; the Octocorallia (Alcyonaria) and the Hexacorallia (Zooantharia). Sea fans, sea whips, sea pens and other soft corals belong to the Octocorallia, while the reef corals, sea anemones, black corals and a number of smaller groups are placed in the Hexacorallia. Reef corals are classified separately in the order Scleractinia (stony corals), which is also known as the Madreporaria.

Corals are not unlike miniature sea anemones encased in a limestone skeleton. Each coral animal is called a polyp. The polyps of one type of Pacific coral are more than one foot in diameter, but in most species they are usually less than an inch wide.

In its simplest form a polyp is a gelatinous tube with a mouth surrounded by a ring of hollow tentacles at the the upper end. The tentacles are unbranched and often end in a bulbous tip. They can be long and conical or short, stubby knobs. Their surface is usually covered with white spots which are dense concentrations of stinging cells known as nematocysts. Reef corals are carnivorous, and they use their nematocysts to catch small planktonic animals. They feed mainly at night when there is a greater concentration of plankton over the reef, whereas during the day many corals withdraw their tentacles into their calcareous skeletons.

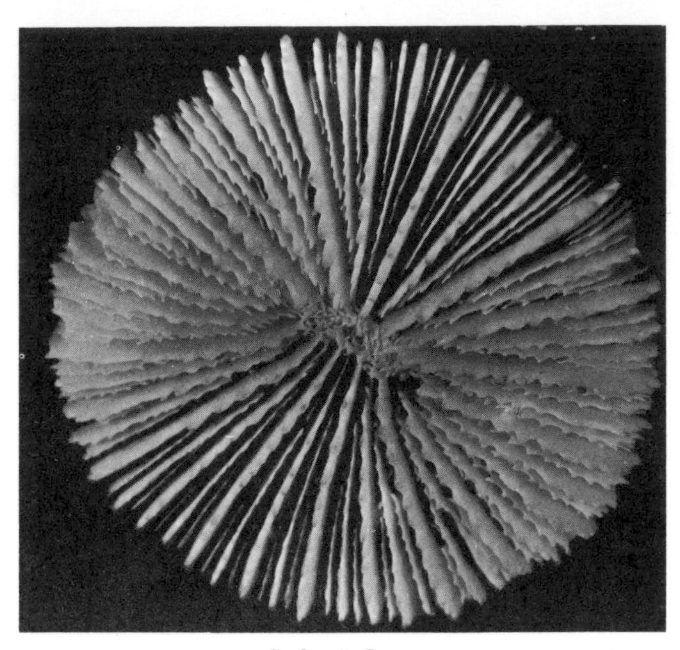

Scolymia Lacera

The skeleton of a coral is secreted by the outer layers of tissue of the polyps. The skeleton of a single polyp is called a corallite or coral cup. Because the polyp is cylindrical, the corallite, at least initially, takes on a cup-like or saucer-like shape. The upper part of a corallite is called a calice. Infoldings of tissue which form indentations on the sides of the polyps secrete additional limestone in the form of radial partitions called septa. The septa converge on the center of the corallite like spokes of a wheel. The upper edge of the septa is sometimes serrated like the blade of a saw. Often the center of the corallite will have a spine-like shaft of limestone, known as a columella, which rises up from the floor.

Each coral starts out as a single polyp forming a single corallite. As the polyp grows larger, it may remain solitary and undivided like many deep-sea corals (and a few reef corals such as *Scolymia* spp.), or it may duplicate itself by a process of asexual reproduction. In the latter case the polyp divides by secreting a wall across the middle of the coral cup. Repeating this process over and over again produces a colony of polyps connected by a common matrix of limestone. In some species new polyps and corallites bud off from the sides of the older corallites. Both processes can lead to the formation of either branching or boulder-like star corals. Asexual reproduction is carried one step further in the brain corals. In these, the original polyp becomes wider and adds more tentacles while more septa appear in the corallite, which changes from being round to oval and then to long and narrow. Instead of dividing in half, the polyp simple adds another mouth at one end and then another and another. As a result, the mouths become arranged in long meandering lines with a row of tentacles on each side. The skeletons of such corals consist of furrows or valleys separated by ridges, with the septa running down the sides of the ridges into the valleys. With their convoluted surfaces and hemispherical shapes, these corals are very aptly called brain corals.

Many colonial reef corals can be many feet thick, but regardless of size, the polyps and living tissues never form more than a thin veneer on the surface of the colony. This is because, as the coral colony grows upward and outward, all the polyps in the colony move up within their corallites while secreting a new floor of limestone above the old one. This growth pattern eventually places the polyps high above their original starting points and setting atop a mass of coral rock which no longer contains living coral tissue.

Although the calcareous skeletons of most species of coral are pure white, living corals are often red, brown, green and yellow, and a few are even blue. The coloration is due to pigments in the tissues of the coral and also due to pigments in the single-celled algae (zooxanthellae) which live in the gastrodermal cells of the coral. There is, however, one shallow-water species *(Tubastrea* sp.) whose skeleton is bright orange, and several deep-sea solitary corals whose skeletons are pale pink.

Corals of a given species can vary considerably in appearance. Differences are due to environmental conditions such as bottom topography, currents, turbidity and surge. Colonies which are exposed to greater wave action tend to be more massive than those growing in quieter waters. Along a narrow and relatively short transect of the reef crest, the elkhorn coral *Acropora palmata* often takes on several different growth forms. This is largely due to progressive changes in the force and direction of the ocean surge.

These variations in the morphology of corals often make identification of species difficult. In some cases it may be difficult to determine if similar corals are different species or only ecological variants of a single species.

Corals first evolved over 500 million years ago. Since that time, thousands of species have appeared and then become extinct. Today, there are over 2500 living species and most of these species are only found in the Indo-Pacific. In the Caribbean there are less than 100 species of shallow-water corals, and only about 20 of these are important reef builders.

STAR CORALS

Some of the major reef builders in the West Indies are the star corals belonging to the genus *Montastrea*. Species of this genus are commonly found on patch reefs in shallow and deep water. The colonies consist of numerous small separate cup-shaped corallites. Species of *Montastrea* are recognizable by the fact that the septa extend above

Star Coral (*Montastrea Cavernosa*)

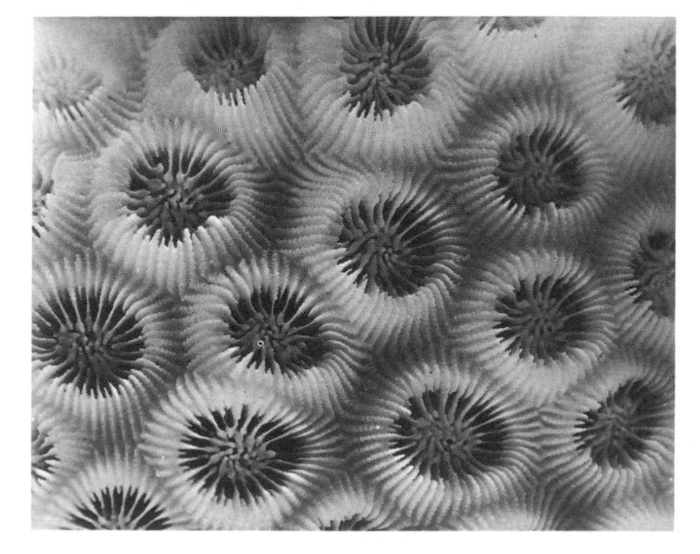

the general coral surface and out beyond the corallite, creating ray-like patterns around each coral cup. In *Montastrea cavernosa* the corallites are about one inch in diameter and about twice as large as those of *Montastrea annularis*. Both species form encrusting or boulder-like masses of coral rock.

Species of the genus *Siderastrea* are also called star corals. *Siderastrea radians* and *Siderastrea siderea* are the most common species of this genus in the Caribbean. In both species the coral cups appear as conical depressions (less steep in *S. radians*) in the surface of the colony. This gives the colonies a pock-marked appearance. Living colonies are yellow tan or maroon red in color. *Siderastrea radians* is found in shallow water, often in areas where gorgonians and sponges are the dominant sessile organisms. This species usually forms small hemispherical colonies rarely more than 1 ft. in diameter. *Siderastrea siderea*, on the other hand, is more common in reef areas where it grows to several times the size of *S. radians*.

SHALLOW-WATER CORALS

In the same shallow-water areas where *S. radians* is found, several other species also occur. One of these is *Favia fragum;* another is *Manicina areolata*. The first species is yellow or brown in color and grows as encrusting or stone-shaped colonies rarely more than several inches in diameter. The coral cups are variable in shape, and quite often elongated (¼ inch wide, and up to ¾ inch long). In *Manicina areolata,* the common rose coral, the corallites are continuous, and thus form long meandering valleys with side branches, as in the brain corals. The entire colony, when viewed from the side, is cone shaped. Small colonies are attached by a stalk at the base of the cone, but larger colonies lie free on the bottom. The largest rose corals are not more than six inches wide, and when alive, they are yellow or brown with greenish valleys and transparent tentacles.

Another species which is common in shallow water, especially in grassy areas, is the tube coral, *Cladocora arbuscula*. This species builds small branching colonies which lie as loose clumps on the bottom. The corallites are about ⅛ inch in diameter and occur only at the ends of the short branches, which are nearly the same width as the coral cups. The fine longitudinal ridges which run down the sides of the branches are actually continuations of the septa.

Tube Coral *(Cladocora Arbuscula)*

POROUS AND FINGER CORALS

Species of the genus *Porites* are important contributors to the structure of the reef. *Porites astreoides* forms multi-lobed encrusting colonies while *Porites porites* and *Porites furcata* (finger corals) build branching colonies. In all the species, circular calices cover the entire skeleton of the coral. The corallites are crowded together and separated by thin incomplete walls, which give the corals a porous appearance when viewed under low magnification. The coral cups are usually less than 1/12 inch in diameter. Both *P. porites* and *P. furcata* have short cylindrical branches, but in the first species the branches characteristically have swollen tips making them appear club-like. In some areas where the water is only a few feet deep *P. porites* may cover extensive areas of the bottom. *P. asteroides,* on the other hand, is more common on patch reefs and in deeper water. It has been reported that major predators of finger corals are the amphinomid worms such as *Hermodice carunculata*. These worms apparently feed on the

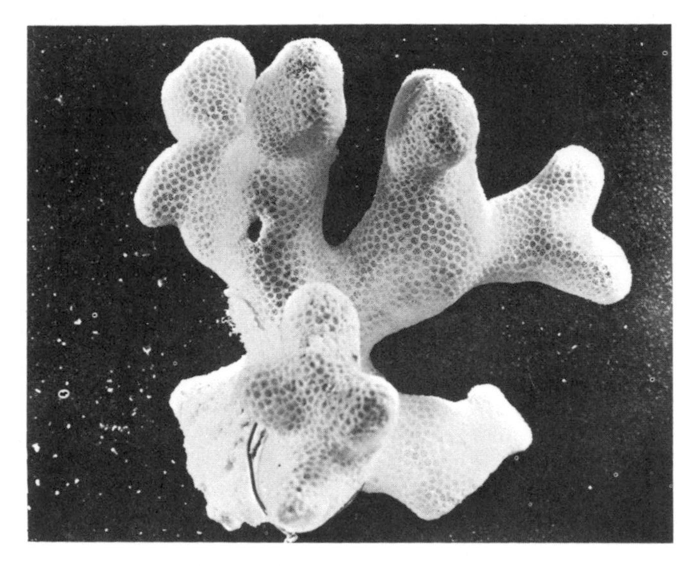

Finger Coral *(Porites Porites)*

polyps near the tips of the branches, leaving only the bare white skeleton.

LEAF AND HAT CORALS

Corals of the genus *Agaricia* are commonly called leaf, hat, or lettuce corals because they form foliaceous colonies. Some grow into large plates or shallow vases which are attached to the substrate by a short stalk, while others build colonies of multiple thin upright blades. The polyps are arranged in parallel rows with low, and sometimes indistinct walls separating one row from another. The septa are very long and continuous between adjacent corallites. The most common species, *Agaricia agaricites,* is partially encrusting in habit but it also develops short upright fronds. This species is common on patch reefs in shallow and deep water. Outside the outer reef different species form huge shingle-like structures which jut out from the wall of the "dropoff." The underside of these colonies is usually heavily encrusted with other marine life.

BRAIN CORALS

There are several genera of brain corals in the Caribbean. Most species are quite common and can easily be recognized in the field. In the genus *Diploria* there are three species, *D. clivosa, D. strigosa* and *D. labyrinthiformis.* In the skeleton of *D. labyrinthiformis* a characteristic feature is a wide groove running along the top of the ridges which separate the valleys. In some colonies this groove may be wider and deeper than the valleys. In the other two species there is no groove or only a very small one, and the septa are very distinct

Lettuce Coral *(Agaricia Agaricites)*

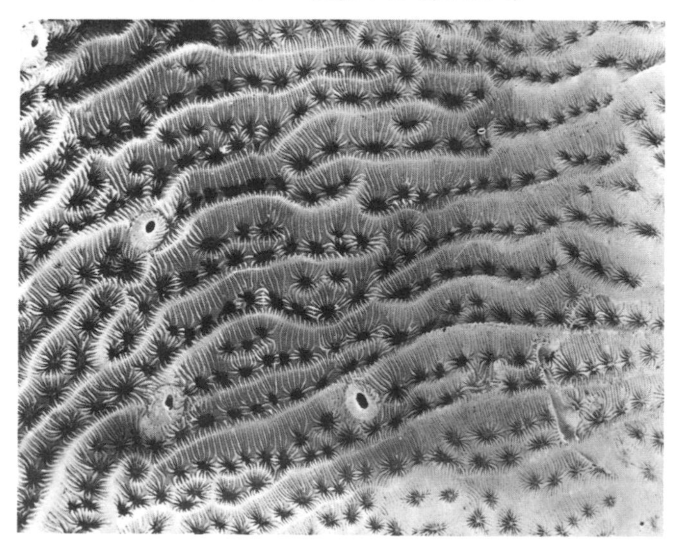

Brain Coral *(Diploria Labyrinthiformis)*

Cactus Coral *(Mycetophyllia Lamarckiana)*

and reach to the top of the ridges. The ridges are very narrow (1/16 inch) and pointed in *D. clivosa,* and wider (¼ inch) and rounded in *D. strigosa.* The latter species forms large dome-shaped masses while *D. clivosa* is usually encrusting.

Colpophyllia is another genus of brain corals, which is classified in the same family as *Diploria.* The skeleton of *C. natans* is so light and porous that it will actually float after being allowed to dry out. *Colpophyllia* differs from *Diploria* in the arrangement of its septa. In *Colpophyllia* individual corallites, with septa radiating out from their centers, can be seen in the floor of the valleys, whereas in *Diploria* the septa are arranged in parallel rows on either side of the valley. Another brain coral which has a similar septal arrangement as *Diploria* is *Meandrina meandrites.* In this species the septa are large, convex, and smooth edged. *M. meandrites* is one of the most common corals on West Indian reefs.

Mycetophyllia lamarckiana is a type of brain coral which is also commonly called a cactus coral. The latter name is derived from the fact that the septa often have large spine-like teeth on their upper edge. This species can also be recognized by the parallel septa which run along the floor of the rather wide valleys and connect the center of one corallite with an adjacent one.

Dendrogyra cylindrus is an unusual type of brain coral in that it is the only one which grows in the form of multi-lobed, pillar-like colonies which rise up several feet from the bottom. This species is usually found in deep water on the outer reef.

ELKHORN AND STAGHORN CORALS

The most important reef building corals in the West Indies belong to the genus *Acropora. Acropora palmata,* the elkhorn coral (see color pictures), is the dominant coral on the edge of the outer reef (reef crest). *Acropora cervicornis,* the staghorn coral, on the other hand, is more common in deeper water, but it may occasionally cover extensive areas of the bottom on the inside of the reef. Because of their branching skeletons, both species are highly susceptible to damage by storm waves. As a result a sizable portion of the coral debris on the reef flat comes from these two species.

Staghorn Coral *(Acropora Cervicornis)*

STINGING CORALS

Stinging corals *(Millepora)* are not true corals. They are actually hydrocorals related to hydroids and jellyfish (Hydrozoa). *Millepora* often grows on the branches of gorgonians, but it can also build independent colonies. One species, *M. alcicornis,* has cylindrical antler-like branches while another, *M. complanata,* forms upright blades. Another hydrocoral, *Stylaster* spp., forms small, delicately branched, pink or purple colonies which are occasionally found under ledges and in caves. Hydrocorals differ from reef corals in that their polyps are very minute and, as a result, the surface of the colony appears smooth.

FLOWER CORALS

The flower corals, *Eusmilia fastigiata* and *Mussa angulosa,* form elaborately branched colonies which are rather easy to recognize. In both species the polyps grow at the ends of the branches. In *E. fastigiata* the coral cups are about 1 inch wide and the edges of the septa are smooth, whereas in *Mussa angulosa* the cups are nearly 2 inches wide and the septa are very spiny.

SOLITARY CORALS

Scolymia lacera is a large solitary coral that is occasionally encountered on deeper parts of the reef. The stony cup of this coral is often more than 3 inches in diameter and the septa are strongly dentate. This species was once thought to be an immature variety of *Mussa angulosa,* but it was later found to be a distinct species. In fact, it has recently been shown that these are two species *(S. lacera* and *S. cubensis)* which differ only slightly in skeletal morphology. However, the species differ considerably in their biology. If the two species settle close together, *S. lacera* will always injure and even kill the *S. cubensis* coral. It does this by means of fine filaments of tissue which extend out from inside the polyp. These filaments are normally used for digestion of food, and apparently they secrete chemicals which damage the tissues of the neighboring coral.

It has also been reported that species of *Scolymia* are not the only kinds of reef corals to exhibit aggressive behavior toward other corals. It seems that there is a hierarchy of aggression among corals whereby those species at the top of hierarchy are always dominant over the others. It has been speculated that this might be a mechanism to protect small corals from being overgrown by large, fast growing species.

CORAL REEFS

Although most corals are restricted to tropical and subtropical parts of the oceans, there are some species which live in temperate and polar waters, and still others which are only found in the deep-sea. Corals of cooler climates and deeper water are usually solitary or only sparsely branched colonies, however there are a few which form reef-like masses of growth. These are found at the bottom of Norwegian fjords, and in deeper parts of the Caribbean and Gulf of Mexico. Generally though, the true reef building corals are only found in warm shallow seas, and there are a number of reasons why they have such a restricted distribution.

Reef corals are called hermatypic corals. This term has come to mean not only that the corals contribute substantially to the rocky substance of the reef, but also that they contain single-celled plants called zooxanthellae. Zooxanthellae are minute cells (dinoflagellates) only 7-10 microns in diameter (a micron is 1/millionth of a meter). They live in special gastrodermal cells of the polyps. This close assocation of animal and plant is an example of symbiosis or "living together."

A number of theories have been proposed to explain this phenomenon. It was originally thought that the zooxanthellae provided oxygen to the polyps while the polyps provided the plant cells carbon dioxide for use in photosynthesis. Today it is considered more likely that the zooxanthellae might provide nutrient substances to the coral while utilizing some of the metabolic wastes released by the polyps. It has also been found that one of the functions of zooxanthellae is to assist in the formation of the coral's calcareous skeleton. Since zooxanthellae require light for photosynthesis they grow best in shallow water above 30 m. Consequently, reef corals also reach their maximum development in the same depths. Although some reef corals do live as deep as 80 m, these colonies usually have less massive skeletons. Below 80 m there are no true reef builders, but so-called ahermatypic species (those lacking zooxanthellae) are found at considerable depths; some as deep as 6000 m.

Because reef building corals cannot survive where water temperatures drop below 70°F, reefs are limited (with a few exceptions such as Bermuda) to the equatorial or subequatorial zones of the oceans. In addition, reefs are only rarely found on the western sides of the continents, and this is also due in part to the low temperatures of the surface waters in these areas. Reef development can also be limited, to some degree, by the upwelling of cold bottom water along the coast, by high turbidity, by unsuitable substrate, or by salinity fluctuations. These latter conditions are often caused by the outflow of freshwater from rivers, and by land runoff.

Although no two coral reefs are exactly alike, the ecological requirements of individual coral species is such that there are some basic patterns of reef growth and development. Where corals grow around an island or along a shore where there is a very narrow shelf, a so-called fringing reef is formed. In this type of reef the line of most active coral growth sits near the edge of a precipitous "drop off" into deeper water. The fringing reef is usually separated from land by a shallow lagoon. On the landward side of the reef is an area of coral debris known as the reef flat. When a reef grows around an island, which afterwards sinks beneath the sea, leaving a circular lagoon, a coral atoll is formed. In the barrier type of reef the outer line of coral growth is many miles from shore and the reef lagoon is often several hundred feet deep.

In the Caribbean there are few atolls and barrier reefs, but many islands have typical fringing reefs while others have bank reefs. In the latter case the corals occur in isolated patches at variable distances away from shore. In some areas there is an outer reef which parallels the shoreline, in addition to patch reefs in more inshore waters. Patch reefs may be surrounded by extensive areas of sand or beds of marine grasses and algae. Where the bottom is hard and covered with only a thin layer of sediment, dense concentrations of gorgonians, sponges and attached algae are commonly found. These gorgonian-sponge communities occur in shallow-water localities throughout the Caribbean.

In areas where the slope of the bottom is gradual, patch reefs may be found near shore while deeper waters may contain stands of gorgonians or beds of marine grasses or algae. Where outcroppings of rock or steep ledges occur in deeper water, the pattern of zonation may be reversed. Thus, it can be seen that the pattern of reef growth in a given locality depends to a great extent on the topography of the bottom. In addition, other environmental factors, such as currents, turbidity and salinity can also play a major role in determining the zonation of reefs, grass beds and gorgonian-sponge communities.

☆

ECOLOGY OF THE REEF

ROBERT BRODY

Marine Ecologist, Island Resources Foundation, St. Thomas, Virgin Islands

Among the biological communities beneath the seas of the West Indies, the coral reef is perhaps the most beautiful. The swaying sea whips and fans, the grotesquely shaped sponges and anemones, the sculptured hard corals and the Walt Disney-looking fishes which comprise the reef community have evolved into a complicated system teeming with life yet delicately balanced. Ecology, as the study of the interactions of living organisms with each other and with their environment, reaches a peak of complexity in the reef system and unraveling the chains of life poses many fascinating problems to the scientist.

The coral reef (and indeed all living communities) is dependent upon two basic factors for its existence: solar energy and chemical nutrients. The levels of nutrients found in the oceans of the western Atlantic around the West Indies are far below those reported nearer the continents and thus the amount of plant life supported by the oceans around West Indian islands is drastically reduced. With few exceptions, only in the shallow island shelf areas adjacent to the islands themselves is the nutrient level high enough to support large numbers of organisms. The bulk of the nutrient material is produced on the islands themselves and carried into the sea by rivers and streams as plant and animal detritus. Acted upon by bacteria and other micro-organisms, this particulate matter is reduced to its chemical constituents; the most important compounds are various organic and inorganic nitrates, phosphates and silicates. This nutrient supply of terrestrial origin is supplemented in some areas by oceanic upwelling which occurs where deep oceanic currents meet or where they are confronted with the physical barriers of island shelf or undersea mountain structures.

Sunlight is of course plentiful in the West Indies, and where this solar energy can combine with a sufficient nutrient supply, the first level of the pyramid of life can begin. The process is called "primary productivity" by ecologists. Since the nutrient supply is greatest in the shallow (less than 100 meters) areas around islands and the solar energy is usually quite high as compared to the temperate zone, the bulk of primary productivity is carried out by bottom dwelling or "benthic" plants. For it is the plants and they alone which can take sunlight and, using their green chlorophyll pigments, convert the sun's energy into the usable carbon-based compounds from which all higher forms of life derive their energy. This concentration of primary productivity in the benthic plants represents a major difference from the ecological system common in the waters nearer the continents. In continental marine ecosystems, a much greater proportion of the primary productivity is done by microscopic planktonic (drifting) plants, probably because of reduced sunlight levels on the bottom and the greater abundance of nutrients in the water over the continental shelf.

What is most peculiar about the coral reef system is that the majority of the "plants" which are doing the primary productivity are either invisible to the naked eye or simply don't look like plants. Some of these plants are brightly hued calcareous algae, red or brown simple plants whose metabolism permits them to extract calcium carbonate from their seawater environment and whose growth form is usually a low, hard mass of limestone. An even larger proportion of the primary productivity is accomplished by plant cells which are microscopic in size and are hidden away among the tissues of hard and soft coral, sea anemone and sponge colonies. These microscopic algae are single plant cells whose habitat is the inside of the coral colony itself. Thousands of these algae cells are present in each coral colony and their cooperative arrangement of living together (termed "symbiosis") is an important example of the kind of close interaction found in the coral

continued on page 83

Descriptions of the following pictures

page 65
THE DROP-OFF
A dramatic view of two divers, one ready for taking photographs, exploring a steep sponge and gorgonian covered drop-off. The divers' air bubbles increase greatly in size as they encounter increasingly less pressure on their rise to the surface. The purplish blue of the depths is clearly visible in this great photograph.

page 66 (upper)
NEON GOBY FEEDING ON PARASITES IN MOUTH OF A GREEN MORAY
A splendid example of symbiosis: This little neon goby (Gobiosoma oceanops) *feasts on parasites in his private dining room, the buccal cavity of a green moray* (Gymnothorax funebris). *The goby is in no danger from the grateful moray.*

page 66 (lower)
REMORAS FEEDING ON A NURSE SHARK
Remoras (Echeneis naucrates) *feed on and clean a nurse shark* (Cynglymostoma cirratum) *at rest in a bed of turtle grass. The remoras do no harm to the shark. The remoras possess a suction disc on top of the head, which they use for attachment to other fish, turtles, and mammals. They often try to attach to a diver.*

page 67 (upper)
TRUMPETFISH HIDING IN BLACK CORAL
A trumpetfish (Aulostomus maculatus) *hides among the branches of an antipatharian* (Antipathes pennacea). *The antipatharians are referred to collectively as black corals, though some species, such as this one, are not truly black. Those that are black are used in the making of jewelry and small carvings.*

page 67 (lower)
SPOTFIN BUTTERFLYFISH WITH PARASITIC ISOPOD
Butterflyfish are some of the most colorful fish of the coral reefs. The black line over the eye is camouflage, which keeps enemies from recognizing and snapping at the eyes. A parasitic isopod clings beneath the eye of this spotfin butterflyfish (Chaetodon ocellatus). *The coral is* Montastrea annularis, *the most massive of the reef builders.*

page 68 (upper)
SARGASSUM FISH
One of the most unusual fish in form, color and design, fully camouflaged here in his floating habitat of Sargassum *weed, the sargassum fish* (Histrio histrio) *is closely related to and somewhat resembles several species of reef-dwelling frogfishes.* Histrio *is a voracious predator and turns cannibal when he has the opportunity.*

page 68 (lower)
LOOK DOWNS
A cubistic portrait of these occasional visitors to the reef. The lookdowns (Selene vomer) *are related to the pompano, permit, and jacks. All belong to the family Carangidae.*

page 69
PORCUPINE FISH
An alarmed porcupinefish (Diodon hystrix) *inflates itself with water. Although this defensive behavior discourages hungry predators, it also greatly reduces its already relatively slow swimming speed. This species is common on some West Indian reefs but ranges as far north as Massachusetts. It has strong jaws with which it crushes the shells of mollusks, sea urchins, and crustaceans.*

page 70 (upper)
WHITE GRUNTS
A fine study of the fast, elegant movements of this beautiful fish, the White Grunt (Haemulon plumieri). *It is a common species, which ranges from Chesapeake Bay to Brazil. In Bermuda it is an introduced species.*

page 70 (lower)
SCHOOL OF BLUESTRIPED GRUNTS
It is highly interesting to note how a school of fish, sometimes numbering by the thousands, keeps organized in a strict order and follows certain

continued on page 81

THE DROP-OFF

NEON GOBY FEEDING ON PARASITES IN MOUTH OF A GREEN MORAY

REMORAS FEEDING ON A NURSE SHARK

TRUMPETFISH HIDING
IN BLACK CORAL

SPOTFIN BUTTERFLYFISH
WITH PARASITIC ISOPOD

SARGASSUM FISH

LOOK DOWNS

THE DROP-OFF

NEON GOBY FEEDING ON PARASITES IN MOUTH OF A GREEN MORAY

REMORAS FEEDING ON A NURSE SHARK

WHITE GRUNTS

SCHOOL OF
BLUESTRIPED GRUNTS

GLASSEYE SNAPPER

SCHOOL OF SQUIRRELFISH ENTERING CAVE

WATCHING STRIPED PORKFISH

ROUGHHEAD BLENNY PEERING FROM HIS HIDEAWAY, A SPONGE

SOUTHERN STINGRAY

YOUNG FRENCH ANGELFISH

MATURE FRENCH ANGELFISH AMONG GORGONIANS

QUEEN ANGELFISH

ROYAL GRAMMA AND PLATE CORAL

DIAMOND BLENNY HIDING
IN GIANT SEA ANEMONE

YELLOWTAIL SNAPPERS
AND SERGEANT MAJOR

FLYING GURNARD

NASSAU GROUPER BETWEEN GORGONIANS

GLASSY SWEEPERS

SUPERMALE OF PRINCESS PARROTFISH

INTERESTING TRIO, LEFT TO RIGHT: SURGEONFISH, PRINCESS PARROTFISH AND REDBAND PARROTFISH

ROSE SPONGE

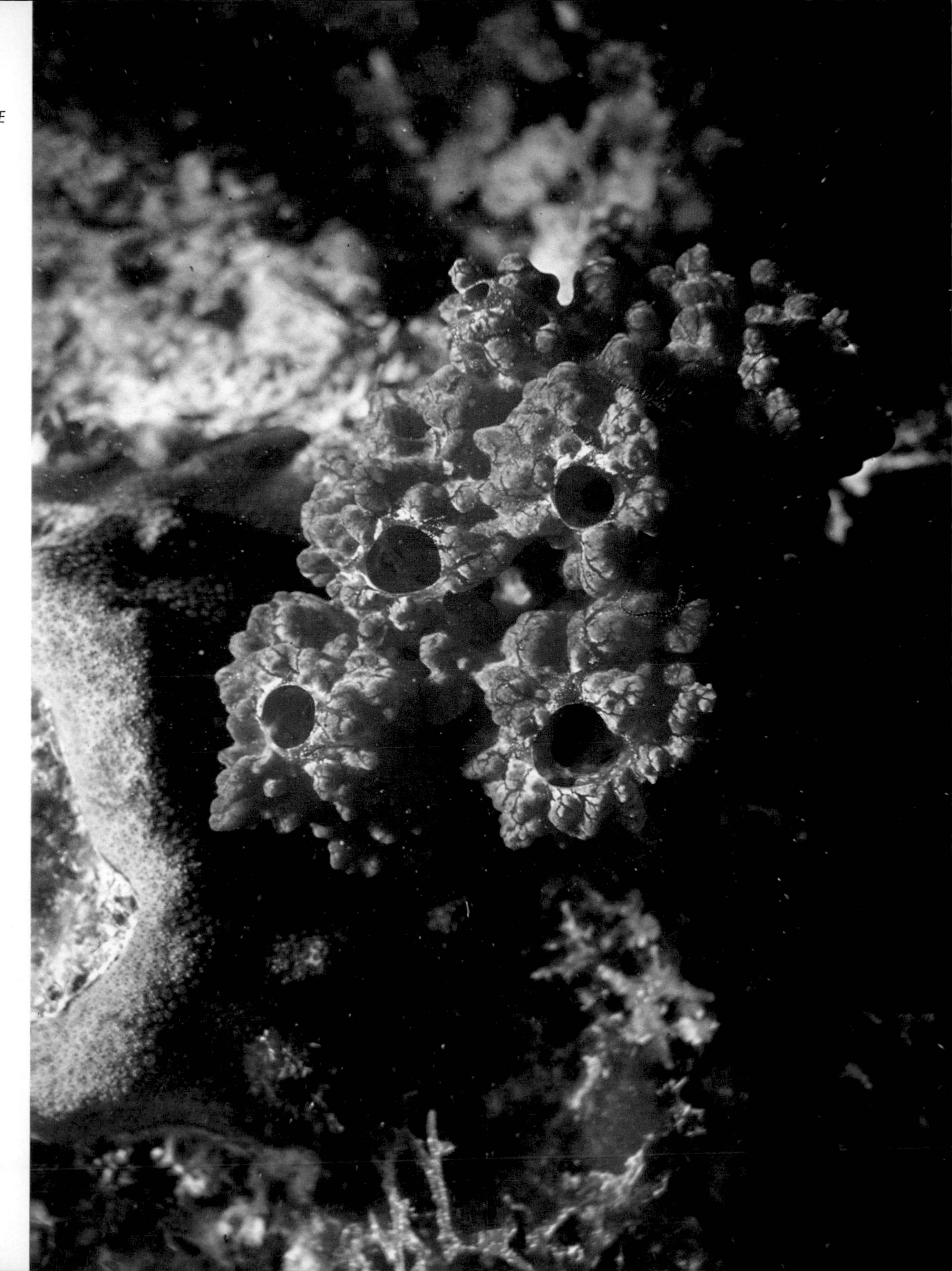

FEATHER STAR EXTENDS FROM FRINGED VASE SPONGE →

REEF COMMUNITY: GORGONIANS, SPONGES, CORAL, SEA URCHIN AND YELLOW-EDGE CHROMIS

HELMET DIVING: EVERYBODY CAN DO IT.

PHOTOGRAPHER IN THE TURNED-OVER WORLD OF A REEF: YELLOW VERONGIA SPONGE AND GORGONIANS GROWING DOWNWARD.

GROUP OF COLORFUL SPONGES

Descriptions of the foregoing pictures

(Continued from page 64)

instincts or leaders in orderly and extended travels in the oceans. The picture shows a big school of bluestriped grunts (Haemulon sciurus).

page 71 (upper)
GLASSEYE SNAPPER
A glasseye snapper (Priacanthus cruentatus) *rests against the sponge and hydroid encrusted framework of an old wreck. The glasseye snapper occurs throughout the tropical oceans of the world.*

page 71 (lower)
SCHOOL OF SQUIRRELFISH
This attractive action photo shows a school of squirrelfish (Holocentrus rufus) *at the entrance to a cave. The small orange corals on the roof of the cave are* Tubastraea tenuilamellosa. *Squirrelfish can be found from Bermuda and Florida through the Bahamas and all over the Caribbean Islands.*

page 72 (upper left)
WATCHING STRIPED PORKFISH
A diver watches a school of striped porkfish (Anisotremus virginicus) *as they swim around a towering mound of corals and sponges. Although the porkfish ranges naturally over a wide area of the tropical western Atlantic, it is an introduced species in Bermuda.*

page 72 (upper right)
ROUGHHEAD BLENNY PEERING FROM HIS HIDEAWAY, A SPONGE
Peering wide-eyed from a hollow sponge, a tiny roughhead blenny (Acanthemblemaria aspera) *seems awe-stricken by the photographer's presence. The purple gorgonian is a young colony of* Pseudopterogorgia.

page 72 (lower)
SOUTHERN STINGRAY
A southern stingray (Dasyatis americana) *glides over the sand inshore from the reef. The barbed stinger on the upper surface of the tail can inflict an extremely painful and slow-healing wound. This is the most common of the stingrays in the West Indian region and ranges from New Jersey to Brazil.*

page 72 (upper left)
MATURE FRENCH ANGELFISH BETWEEN GORGONIANS
A French angelfish (Pomacanthus paru) *glides among gorgonians. The plumelike gorgonians* (Pseudopterogorgia *sp.*) *in the foreground are commonly called sea feathers.*

page 73 (upper right)
YOUNG FRENCH ANGELFISH
A young French angelfish (Pomacanthus paru) *examines the surface of a rocky ledge which supports several colonies of plexaurid gorgonians. The yellow bands will disappear as the fish matures and develops the color pattern of the adult as shown in the adjoining photograph.*

page 73 (lower)
QUEEN ANGELFISH
One of the most colorful and most beautiful fishes of the coral reef is the queen angelfish (Holacanthus ciliaris), *easy to recognize by the dark dot on the forehead. Changing color with reflections and every turn, it is fairly common in the Caribbean, Bahamas, Bermuda and Florida.*

page 74 (upper left)
DIAMOND BLENNY HIDING IN GIANT ANEMONE
A diamond blenny (Malacoctenus boehlkei) *hides among the tentacles of a giant anemone* (Condylactis gigantea), *which would be dangerous for other fish. There is a quiet understanding between these two for mutual assistance. Notice the fine details of the tentacles of the anemone, so well reproduced in this photograph.*

page 74 (upper right)
ROYAL GRAMMA AND PLATE CORAL
The brilliant colors of the royal gramma (Gramma loreto) *stand out against the undulating surface of the plate or hat coral* (Agaricia fragilis). *The royal gramma, growing to little more than three inches, is also called the fairy basslet.*

page 74 (lower left)
YELLOWTAIL SNAPPERS AND SERGEANT MAJOR
Yellowtail snappers (Ocyurus chrysurus) *and a sergeant major* (Abudefduf saxatilis) *congregate on a section of reef that appears to be dying. The coral head is partially covered with sediment and algae overgrowth. In the background, portions of a sea feather have died, leaving the horny axis exposed.*

page 74 (lower right)
FLYING GURNARD
In spite of its winglike fins, the flying gurnard (Dactylopterus volitans) *is essentially a bottom dweller. It is a native of the waters of both sides of the Atlantic and grows to be up to one foot in length. In the western Atlantic it occurs from Bermuda and Massachusetts southward to Argentina. It sometimes uses its pelvic fins for walking on the bottom.*

page 75
NASSAU GROUPER BETWEEN GORGONIANS
A Nassau grouper (Epinephelus striatus) *cruises between sea feathers (above) and plexaurid gorgonians below. This grouper is very variable in colors, changing ground colors from white to brownish, greenish, grayish and bluish tones. The irregular dark stripes and markings are mostly dark olive or brown. Nassau groupers can be found from Bermuda to the Bahamas and Florida and South to Brazil.*

page 76
GLASSY SWEEPERS
A tunnel in the coral rock is filled with woeful-looking glassy sweepers (Pempheris schomburgki). *The picture shows mature specimens with the typical coppery color. Young sweepers are almost transparent. They live around Bermuda, the Bahamas, Florida and the Caribbean, preferably in coral caves.*

page 76 (upper right)
SUPERMALE OF PRINCESS PARROTFISH
This handsomely colored fish is a supermale of the princess parrotfish (Scarus taeniopterus), *described in lower picture on this page.*

page 76 (lower)
INTERESTING TRIO: SURGEONFISH, PRINCESS PARROTFISH AND REDBAND PARROTFISH
A lavender sea-feather (Pseudopterogorgia acerosa) *waves before a mixed threesome. In the right background is a redband parrotfish* (Sparisoma aurofrenatum), *and in the left foreground is an ocean surgeonfish* (Acanthurus bahianus). *In the center of the photograph is a princess parrotfish* (Scarus taeniopterus). *Immature princess parrotfish, normal mature males, and mature females all possess the color pattern seen here. Astonishingly, some normal females transform into beautiful supermales such as the individual shown in the photograph, upper right of this page. Princess parrotfish are found around Bermuda, the Bahamas and throughout the Caribbean Islands.*

page 77
ROSE SPONGE
A view directly down into the oscules of an attractive vermilion-colored sponge. On the surface, numerous radial canals can be seen leading from the oscules.

page 78 (upper)
A REEF COMMUNITY: GORGONIANS, SPONGES, CORAL, SEA URCHIN AND CHROMIS
Yellow-edge chromis (Chromis multilineata) *investigate a colorful sponge and coral community. The spines of the black sea urchin* (Diadema antillarum) *project from beneath an iridescent pink and blue sponge* (Callyspongia plicifera). *The coral in the background is* Montastrea cavernosa.

page 78 (lower)
HELMET DIVING: EVERYBODY CAN DO IT
One of the interesting diving attractions in Nassau in the Bahamas in the winter and in Bermuda in the summer is helmet diving with the Hartleys into the underwater wonderland around those islands. Everybody can do it: non-swimmers, those who wear glasses, people of all ages. No lessons are needed. Ladies, the helmet keeps your hair dry. And what an experience, to see living coral, colorful reef fishes and sponges; to walk on the

ocean floor in the clearest warm water in the world. The reef in the picture is composed of scattered corals, sponges and gorgonians. The big basket sponge in front of the suited diver is Ircinia campana, *one of the most common sponges in West Indian waters.*

page 79

FEATHER STAR EXTENDS FROM FRINGED VASE SPONGE

A crinoid or feather star (Nemaster rubiginosa) *extends its plankton ensnaring arms from the hollow of a fringed vase sponge* (Dasychalina cyanthina). *The lavender-blue color of the sponge is obscured in some places by the abundant polyps of a zoanthid* (Parazoanthus *sp.*).

page 80 (upper)

PHOTOGRAPHER IN THE TURNED-OVER WORLD OF A REEF

An underwater photographer examines the topsy-turvy world of the underside of a ledge. The large yellow sponge (Verongia *sp.*) *and the gorgonians* (Iciligorgia schrammi) *are growing in a direction completely opposed to their normal orientation.*

page 80 (lower)

GROUP OF COLORFUL SPONGES

This photograph shows an interesting group of three different sponge species, growing together and in each other, giving the appearance of one single body.

☆

continued from page 63

ecosystem. The reef building or "hermatypic" corals are limited to the tropics by warm temperatures found there, but more important are limited to the shallow areas of the tropics where sunlight can penetrate with sufficient intensity to carry out photosynthesis (the "euphotic zone") by their symbiotic algae. This symbiosis can be so close, so important, that some species of soft corals have been termed "obligate phototrophs" because they simply cannot live without the sunlight which allows their algal symbionts to photosynthesize. The symbiotic algae may contain chlorophyll and be called "zoochlorellae" or both chlorophyll pigments and brown pigments called carotenoids or xanthophylls and the algae are thus termed "zooxanthellae."

The coral reef is a community of organisms. Its most prominent members and its literal "backbone" are the corals themselves but these colonial animals are by no means the only important residents of the reef. The various grasses and algae at the base of the reef combine with other plants to produce the food of a myriad of animals. The invertebrate animals on the reef include representatives from virtually every group found in the sea from lowly protozoans and other microscopic forms to the complex crabs, molluscs and sea worms.

The corals and their relatives of the group termed coelenterates or cnidarians by biologists are the most conspicuous of the animals attached to the sea floor on the reef. The group are all close relatives of the jellyfish; this class of animals includes the stony corals, the sea fans and sea whips or gorgonians and the sea anemones. Despite their plant like appearance, these are animals who live as a colony spread over the bottom (corals and anemones) or as branching tree-like forms (gorgonians). All have small (but still visible) tentacles and a variety of stinging cells (nematocysts) in common with their jelly fish relatives.

While most corals and gorgonians have innocuous nematocysts, woe to the unwary snorkeler who brushes against one of the mustard colored "rocks" that is a fire coral or a feathery relative, the hydroid colony. The sensation has been likened to touching a hot flat iron by many and has produced shock and even respiratory paralysis in some sensitive individuals.

The coelenterates are considered quite primitive (in an evolutionary sense) because they have only two distinct embryologic tissue types whereas most higher animals (including mammals) have three. The coelenterates use this difference in a profitable way by forming complex colonies of hundreds or thousands of individual animals of two tissue types which can transfer food and waste products amongst themselves by amoeba-like cells which move between the individuals in the colony. The coral colony releases reproductive cells directly into the surrounding sea water where fertilization takes place, then a drifting (planktonic) larval stage is formed which eventually settles to the bottom. If the bottom is suitable a new colony starts; all of the hundreds of animals in a coral colony are formed by asexual reproduction from the original single larval one. Only a tiny fraction of the coral larvae are successful in surviving their planktonic voyage and then settling to start new colonies.

This dispersion of larvae by planktonic forms and the colonial growth form is shared by another important group of reef animals, the sponges. Termed Porifera by biologists and even less complex than the corals, sponges are colonies of single celled animals who somehow are always capable of producing a colony of growth form specific to their species. It has been demonstrated that many species of sponges can be broken up and strained through cheesecloth yet will always reunite in the form typical of their original colony frequently forming several smaller colonies.

The more complex animal species found on the coral reef are largely capable of mobility and as such may also be found in many other habitats beneath West Indian waters. The starfishes, brittle and basket stars, sea cucumbers and sea urchins (the echinoderms in biological nomenclature) all share the common feature of "pentaradial symmetry" where the organ systems and general growth form is symmetrically arranged in five radii around a central axis through the mouth and anus rather than the more common "bilateral symmetry" found in most other multicellular forms (such as man). This pentaradial symmetry is obvious in the star shaped forms and can be seen upon the shell of a sea urchin or by looking closely at the "end" of a sea cucumber. The starfish, urchins and cucumbers move on "tube feet," small extensions from the body, which are powered by changes in pressure in a "water vascular system," a semi-closed circuit of canals and vessels with the fluid moved by microscopic hair-like structures called cilia and by muscles outside the vessels. The brittle and basket stars take this system one step further and are actually able to push and pull their way through the network of attached organisms growing on the reef or get up and scamper along the bottom using their whole arms. The tube foot means of locomotion might seem slow to the untrained, but if a snorkeler were to find an aggregation of the long spined black sea urchin *Diadema* and break one open, he would be surprised to see how tube feet and spines combine to scatter the group in a big hurry. Incidentally, *Diadema* is reported by diving instructors and underwater tour guides to be probably the most dangerous single animal on the reef; thousands of visitors to the West Indies are painfully hurt by their spines every month while unprovoked shark and barracuda attacks are probably limited to less than one in any given year. The experiment outlined above should thus be undertaken with great caution.

The biological group known as molluscs include the familiar snail and clam shaped seashells and the less obviously related squids, octopods, sea slugs and chitons. All of these animals possess a glandular structure (the mantle) which is capable of extracting calcium carbonate (limestone) which is dissolved in sea water and depositing it as a hard compound. This material is quite obvious in the shells of snails, clams and chitons but the squids have their "shells" inside the body wall. The economically important molluscs of West Indian reefs are not the pearl-bearing oysters but the edible conch and whelk species. A number of gastropod and bivalve forms are valuable to shell collectors but ecologically perhaps the most important molluscs are the boring clams and gastropods who make their home below the living surface of the coral reef. They bore through the coral rocks, recycling the hard carbonates in the system and, upon their death, providing habitats for a number of smaller forms.

All of the groups mentioned above (cnidarians, sponges, echinoderms and molluscs) are reef builders in the sense that their natural processes permit them to extract dissolved chemicals from their sea water environment and deposit it as hard mineral compounds. Much of this material is calcium carbonate which is indeed the building material of the reef. The stony corals make an obvious contribution with their hard skeletal remains as do the molluscs with their shells. A new layer of carbonate is constantly being laid down and recent research has shown that distinct animal growth bands can be detected in x-ray photographs of sections cut from coral colonies. The gorgonians and some sponges make their contributions by secreting microscopic rods of calcium carbonate which are held together by the living tissues of the colonies. When the colony dies, its carbonate components are released to become a part of the non-living sediments on the reef. The gorgonians alone have been estimated to contribute more than one ton per acre on reefs in the Florida Keys. Many of the sponges found on the coral reefs produce microscopic rods of silicates which are extremely sharp glass-like needles; the untrained diver should be extremely cautious in handling *all* sponges. The echinoderms all have skeletal elements composed of calcium carbonate; the most obvious is the cushion-shaped shell or "test" of the sea urchins but when dried, the skeletons of starfish or brittle stars display their thousands of carbonate plates. Some of these smaller calcium carbonate particles are swept away by currents or dissolved but a substantial amount must be incorporated into the reef.

Besides producing coral reefs, the calcium carbonate sediments produced by reef animals and calcareous algae are of great economic importance in many islands of the West Indies. Lacking extensive fresh water rivers, these islands are almost totally dependent upon calcium carbonate producing organisms for the sand on their beautiful beaches and, as important, for the sand in the concrete used in construction. From a conservation point of view it is ironic that the massive development of tourism in many West Indian islands requires this sand and yet wantonly destroys coral reefs by dredging, filling and pumping sewage and other pollutants into the sea. A good example of damage to coral reefs caused by dredging and filling is now becoming obvious in southern Florida. Florida, however, is part of the continent and has vast reserves of construction aggregates which are unavailable (except at prohibitive cost) to the West Indies.

The remaining prominent group of invertebrates of the coral reef is the Crustacea, a part of the largest phylum of invertebrates, the arthropods. There are no insects on the coral reef but there are several hundred species of crabs, lobsters and microscopic crustaceans who are full time residents. The microscopic forms are far too numerous to describe here but their place in the reef community is nonetheless important. They are largely plant and detritus feeders who function as the link in the food chain which transfers vegetable matter and the break-down products of animal remains to the animal members of the community. They find their way into the digestive systems of members of virtually every invertebrate and vertebrate group on the reef. The crabs, shrimp and lobsters on the reef are generally scavengers, cleaning up the remains of dead organisms on the reef but also actively feeding on many of the attached animals. The lobsters include two species of important commercial value. West Indian lobsters are sold throughout the region and exported as rock lobster tails. The shrimps of the coral reef are ecologically important as scavengers and also as parasite removers to many fish. The red and white banded shrimp *Stenopus* can frequently be seen hovering around small fish who float motionless to have their external parasites removed by the shrimp. The edible shrimp species so important to commerce are creatures of the shallow marine grass beds and not abundant in the West Indies.

The fishes form the last group of reef organisms which we will discuss here. An entire volume could be devoted to reef fish and their ecology (and indeed many have). We shall be brief, mentioning only those fish which have direct significance to the groups of organisms already mentioned. The fish food chain proceeds in a logical manner with the smaller species generally feeding on even smaller organisms and the larger species feeding in turn upon smaller fish. The largest carnivores on the reef are strictly fish-eaters with barracuda, grouper and snappers most prominent among them. They are augmented by the roving jacks, mackerals, and sharks who pass through the reef area on their seemingly endless migrations. Many smaller fish species are strictly plant and detritus feeders while some are less specialized and eat molluscs, crustaceans and other small invertebrates. A number of fishes are adapted to feeding directly on the corals, sponges or upon smaller forms inhabiting the coral rock and as such are important contributors to the sediment supply on or near the reef. It has been estimated that reef fishes might excrete as much as one ton per acre per year of fine calcareous sand after digesting the available organic material. When a coral reef is observed from the air a band of white is almost always noticeable dividing the active coral growth from the grasses and algae which are usually present at the reef's base. This band is probably maintained by the constant nocturnal browsing of herbivorous fishes who hide in the crevices of the coral during the day and by the fine calcareous sediment excreted by the coral feeders.

Of course the abundance of fish life on the coral reef provides an important food source for man but this bounty is not without its problems. A number of kinds of poisoning are attributed to reef dwelling fishes and in several places in the West Indies, fish poisoning is a severe public health problem. The most common type of poisoning is known as ciguatera or barracuda poisoning. This intoxication affects individuals who have eaten fish which may be perfectly fresh and properly prepared and which contain a toxin (or perhaps toxins) which the fish has acquired in his environment. Ciguatera poisoning has been reported in the West Indies since Columbus' time and thus is probably not a result of man's pollution of the sea. The most likely origin of ciguatera toxins is a small, perhaps microscopic plant or fungus which is only toxic in certain highly localized regions or which only becomes toxic when influenced by certain unknown factors in the environment. While ciguatera is usually not a fatal disease, one or two deaths are reported on the average every year from the Virgin Islands-Leeward Islands area. In the Virgin group alone there are as many as 50 cases each year who seek medical assistance or about 1 case reported for each 2000 residents. Most frequently the poisoning results in severe gastrointestinal symptoms followed by weakness and muscle or joint pain for several days or weeks. Moderately severe cases also show a variety of neurological symptoms including a peculiar "Paradoxical Sensory Disturbance" wherein the victim feels that his sense of hot and cold is reversed; victims have been observed to blow on their ice cream for as long as several weeks after the poisoning incident.

The most frequently implicated fish are the large specimens of barracuda, grouper, jack and snapper but by no means are all of the fish of a given species toxic. Visitors to the West Indies are advised to consult local physicians, public health officials or fishermen before buying fish; most fish available in markets can probably be considered safe. The only known tests for ciguatera toxin involve feeding the fish to a cat or mongoose or require sophisticated laboratory preparations; none of the West Indian folklore involving silver, sweet potatoes, flies not landing on the fish, etc. have any reliable validity.

Not everything involving human illness and the coral reef is as negative as fish poisoning. Recent research has shown that a variety of West Indian reef organisms contain biologically active compounds which may be of great use in curing a number of diseases. In many cases pharmaceutical research has been able to obtain the natural compound in extracts from corals, gorgonians and sponges. Some of the compounds reported to date have shown excellent anti-bacterial, anti-fungal and in some cases, anti-cancer activity. Wherever preliminary testing has shown this antibiotic activity, research chemists make an effort to isolate, characterize and then synthesize the active compound. This might lead to production of drug substances in factories hundreds of miles from the coral reef where the compound was discovered.

In at least one case, synthesis of a pharmaceutical from coral reef sources has been extremely difficult and research is now underway to determine the feasibility of harvesting the natural material in a true sea-farming system. The "once a month" birth control pill recently developed is based upon compounds called prostaglandins. These compounds are found in considerable quantities in a relatively common reef gorgonian named *Plexaura homomalla.* Perhaps the future of West Indian coral reefs may include neatly planted rows of *Plexaura* with divers harvesting the drug-bearing colonies in commercially valuable quantities. *Plexaura,* incidentally, is only found in the tropical seas of the Western Atlantic and is most abundant in the West Indies.

Just what a gorgonian does with a mammalian hormone-stimulating chemical is still a mystery. Like many of the complexities of the coral reef system this question remains and the future should hold some of the answers. Knowledge of the ecology of the coral reef community in the West Indies is a story of many more questions than answers.

☆

GORGONIANS

DENNIS M. OPRESKO

Museum of Cooperative Zoology, Harvard University

Of all the colorful varieties of coral which flourish in the warm shallow waters of the Caribbean, Bahamas, Florida and Bermuda, probably the most exotic are the gorgonians. These are the plant-like seafans and seawhips which form, in some areas, vast underwater coral "gardens." With their multi-colored and elaborately branched colonies which sway and bend with the motion of the sea, gorgonians lend to West Indian reefs a very distinctive flavor which is almost entirely lacking on reefs of the Pacific.

Gorgonians are distant relatives of stony reef corals, and like most reef corals they are colonial animals with thousands of individual anemone-like polyps covering the surface of each colony. They differ from reef corals in that they do not produce massive calcareous reefs, but build, instead, simple or highly branched, flexible "skeletons" made of a horn-like substance called gorgonin. In color and consistency gorgonin is often remarkably similar to wood. Thus it is not unusual for a beachcomber to mistake a bare beachworn skeleton of a gorgonian for some strange, leafless marine plant. From the time of Aristotle until well into the 18th century, naturalists and laymen alike were also misled by the tree-like appearance of gorgonians, and when a scientist of the 17th century made a chemical analysis of a gorgonian and produced a volatile substance with an odor not unlike that of cabbage, this was seen as conclusive proof that gorgonians were really plants. But even from the earliest observations of the ancient Greeks there was some suspicion that the polyps were not really the counterparts of the flowers of terrestrial plants. Some naturalists thought that the polyps were wormlike animals that simply inhabited the "bark" of the coral "trees." This dual nature of gorgonian corals led to their being classified as zoophytes, a word derived from the Greek meaning animal-plants. Other colonial marine animals such as reef corals, sponges, bryozoans (moss animals), and even some marine plants, were also described as "zoophytes." Eventually careful studies of living specimens revealed the true animal nature of many of these zoophytes. Gorgonians have since been classified, in the animal phylum Coelenterata, together with the reef corals, jellyfish, sea anemones, and hydroids.

One of the more plantlike characteristics of many coelenterates is that they are sessile; that is to say, they are permanently or semi-permanently attached to or embedded in the sea bottom. On land there are few, if any, sessile animals, but in the sea conditions are often ideal for animals to adopt this type of life style. In the sea there is an abundance of microscopic plants and animals called plankton. The small crustaceans, worms, snails, fish, and other minute organisms which make up marine plankton live out part or all of their life cycle passively drifting through the ocean. Sessile animals are specially adapted to feed on plankton. Thus they need not move about in search of food; they simply wait for currents to bring the plankton to them. In addition to coelenterates, barnacles, sponges, tunicates (sea squirts), bryozoans (moss animals) and plume worms are just a few of the many other kinds of marine organisms which have adopted a sessile way of life.

Although many gorgonians are known to be active plankton feeders, it has been postulated that they may also satisfy part of their nutritional requirements by extracting dissolved and particulate organic matter from seawater. Another theory suggests that the symbiotic algae which live in the tissues of the gorgonians may release organic substances which are utilized by the corals. The algae are single-celled dinoflagellates called zooxanthellae. As plants they can convert simple inorganic compounds, such as carbon dioxide, into complex organics. The exact function of the zooxanthellae in relation to the physiology of the gorgonians is poorly understood, but it has been reported that some soft corals will survive without food as long as they possess their zooxanthellae, but that they will starve to death in the presence of excess food if they are deprived of their algal symbionts.

As sessile plankton feeders, most gorgonians have evolved into colonial organisms with large branching skeletons. A colonial organism such as a gorgonian has the capacity to reproduce polyps asexually by budding. The large number of feeding polyps which are formed in this way contribute to the production and maintenance of the large supporting skeleton. The skeleton, in turn, serves several functions. It allows a coral to be fastened securely to the bottom, protects it in a limited way from predation and siltation, and also places the feeding polyps high above the bottom in an advantageous position for collecting plankton. Gorgonians are thus well adapted for their niche in the marine community.

Although gorgonians are found in all the seas of the world from the Arctic to the Antarctic, they are most abundant in the warm shallow waters of the tropics. Over seventy species occur in the Caribbean and Bahamas, and seventeen of these tropical species also occur as far north as Bermuda. Within the Caribbean, gorgonians are most abundant in reef areas where the bottom is hard and the water clear. Almost all of the species on the shallow reefs belong to either the family Gorgoniidae or to the family Plexauridae. On deeper reefs, below 100 ft. or so, a different gorgonian fauna exists. The predominant families here are the Paramuriceidae and the Ellisellidae. Species of these families range down to a depth of several hundred fathoms or more where they are replaced by the families Primnoidae, Chrysogorgiidae and Isididae. The latter families are true deep-sea gorgonians. They often have very extensive vertical ranges and some have been collected at depths greater than 12,000 ft.

The gorgonians of the shallow coral reefs are not distributed as haphazardly as one might conclude from casual observations. Since each species has its own set of tolerances for such variable environmental factors as temperature, salinity, siltation, currents, and surge, each is most abundant in areas where conditions are most favorable. The reef environment, however, is usually never so severe as to exclude species from less favorable areas. Consequently, at most localities there is usually a great variety of species. On some patch reefs, there may be as many as thirty different species represented, and a single acre of bottom may contain tens of thousands of colonies.

SEA FANS

Of all the gorgonians of the reef, probably the best known and most beautiful are the sea fans. These large purple and yellow, fan-shaped colonies, with their pastel coloration and delicate latticework of branches, add a touch of elegance to the seascape. Actually there are several different kinds of sea fans, but all belong to the genus *Gorgonia*. As one might suspect, it is from this scientific name that the word gorgonian originates. Sea fans were originally named after the "Gorgons" of Greek mythology who had snakes for hair and eyes which, if looked into, turned the beholder into stone.

Gorgonia ventalina, with its more distinctively flattened branches, is found over wide areas of the reef. *Gorgonia flabellum*, however, is often restricted to very shallow water where there is a very strong surge. The colonies tend to be orientated in the same direction, perpendicular to the current. The great flexibility of their skeletons allows them to withstand the force of the sea by bending but not breaking.

SEA FEATHERS AND SEA PLUMES

The numerous purple sea plumes and sea feathers which are found on the reef belong to the genus *Pseudopterogorgia*. There are about twelve different species and most are recognizable by their pinnately branched colonies. The most common species are *Pseudopterogorgia americana, P. acerosa* and *P. bipinnata*. The first two species are commonly found together in rather shallow water. Both produce large plumose colonies with long, slender, drooping branchlets. *P. americana* can be distinguished from *P. acerosa* by its slimy surface and by the tendency of its polyps to remain expanded. *P. bipinnata* is more common on patch reefs and in deeper water. It produces a more regularly branched colony with shorter and stiffer branchlets.

Although species of the genus *Pseudopterogorgia* are superficially very unlike the sea fans, they are both placed in the same family Gorgoniidae. This is because they both have very similar microscopic calcareous spicules embedded in their tissues. Spicules are a key character in classifying gorgonians, and to accurately identify any species, one must extract the spicules from the surrounding tissue and examine them under a microscope.

Another group of species belonging to the genus *Pterogorgia* is also placed in the family Gorgoniidae. These species form small shrub-like colonies which are found in the same habitats occupied by *Pseudopterogorgia americana* and *Pseudopterogorgia acerosa*. The colonies are variable in color and can be purple, gray, green or yellow. In *Pterogorgia citrina* the branches are flattened and the polyps are confined to rows along the narrow edges. The branches of *Pterogorgia anceps* are also flattened, but near the base of the colony they are often three or four-sided.

PLEXAURIDAE

The second large family of shallow-water gorgonians is the Plexauridae. In this family there are about seven genera. As in the case of the Gorgoniidae, these genera are grouped together on the basis of spicular morphology and not because of similarities in the growth form of the colonies. Most plexaurids have a shrub-like type of colony, and this makes identifications in the field almost impossible. However, there are some basic external characters by which some genera and a few of the more common species can be recognized. Generally though, the colonies have a variable number of straight or slightly curved, upright branches, and the branching pattern can be dense and bushy or thin and scraggly. The color is usually a dull grey or tan or a dark brown or black.

One of the more common plexaurids on patch reefs is *Plexaura homomalla*. This species produces relatively concentrated amounts of prostaglandin chemicals which have many important pharmaceutical uses. The demand for these chemicals is so great that there may come a day when *P. homomalla* is farmed on the sea bottom like fruits and vegetables are on land. This species is tan in color, densely branched, and can grow to a height of more than three feet. It can be recognized by the tendency of many of its smaller branches to grow out from only the upper side of the larger branches. *Plexaura flexuosa* is often similar in appearance, but occasionally red in color.

In many species of the plexaurid genus *Eunicea* the surface of the colony is distinctly rough and bumpy. This is due to the presence of large tube-like calyces which surround the openings of the polyps. The colonies are generally irregularly branched, but one species, *Eunicea tourneforti*, characteristically grows in the form of a living candelabrum.

In the genus *Pseudoplexaura*, the surface of the colonies is smooth, but the polyp openings are usually very crowded together giving the colony a very porous appearance. Most species are sparsely branched with only a few long cylindrical branches; hence the appropriate common name "seawhips."

One character that almost all species of *Plexaurella* have in common is the tendency of the polyp openings to be reduced to mere slits in the surface of the colony. There are six species in the genus and some have a rather wide distribution. They can be found in shallow water, on patch reefs, and also in the deeper waters of the outer reef. Most have very thick branches and large polyps, and some can grow to a height of four feet or more.

Species of the plexaurid genus *Muricea* can easily be recognized by the spiny surface of the colonies. This is caused by large thorn-like spicules which are heavily concentrated below the polyp openings. The spicules are amber colored and thus give a yellow or orange cast to the entire colony.

OTHER GORGONIANS

An unusual gorgonian found on patch reefs and in deeper water is *Erythropodium* sp. The colonies, purple in color and covered with very tall brownish polyps, usually grow over rocks, dead gorgonian branches, or other bottom debris. *Briareum asbestinum* is very similar in appearance, but colonies of this species often send up short stubby, finger-like stems. The center or core of these stems does not contain the woody axial material characteristic of most gorgonians. Instead it contains a mass of densely packed calcareous spicules. Consequently the upright colonies of *Briareum* are rather rigid. On deeper reefs below 90 or 100 ft., there is a large multi-branched gorgonian called *Iciligorgia schrammi*. This species is similar to *Briareum* in that it has the same kind of stiff calcareous axis. Fortunately, in the calm waters of the deep reef it is not susceptible to the damage it might suffer in the more turbulent waters of the shallow reef. Conversely, it can be seen how the sea fans and sea whips, with their flexible skeletons of gorgonin, are better adapted for life in shallower waters.

GUESTS, FRIENDS AND ENEMIES

Many different kinds of marine organisms can be found living on or about the branches of gorgonians; some use the gorgonian skeleton as a substrate on which to settle and grow, others hide among the branches, and still others feed on the living polyps. The surface of the branches, especially if they have been denuded of their covering of living tissue, is an ideal substrate for the attachment of encrusting organisms such as tunicates (sea squirts), bryozoans (moss animals), hydroids, stinging coral *(Millepora)*, and various types of algae and sponges. Several species of oysters also attach themselves to gorgonians. The winged oyster *(Pteria* spp.*)* uses its byssus threads to secure itself to a branch, while the leaf oyster *(Ostrea frons)* has modified part of its shell into finger-like projections which grow around and clasp a branch. Both bivalves feed on planktonic organisms which they filter out of the surrounding water.

One of the more spectacular residents of gorgonian colonies is the basketstar *Astrophyton muricatum*. This creature is a distant relative of starfish and like starfish it has five arms extending out from its flat, disk-like body. In a basketstar the arms are very long and elaborately branched and when fully expanded an entire animal can be more than several feet in diameter. During the day skindivers rarely notice an *Astrophyton* on the reef; at night, however, it often seems that every gorgonian colony has one or more basketstars clinging to it. The basketstars perch themselves on the branches of a colony and spread their arms into the current to feed on plankton. With the coming of dawn, they coil up and rest in the lowermost branches of the gorgonian or retreat into crevices in the reef where they are safe from predators.

Basketstars apparently do little harm to their gorgonian hosts; however, there are a number of marine animals which prey specifically on gorgonians. Perhaps the most famous of these is the flamingo tongue shell, *Cyphoma gibbosum*. With its bright orange, black and white mantle it is very conspicuous on the purple and yellow sea fans and sea plumes on which it lives and feeds. Trails of denuded axis are mute evidence that tongue shells are at work, and when several tongue shells inhabit the same colony, the gorgonian can suffer considerable damage.

Snails of the genus *Neosimnia* also prey on gorgonians. The shells of these gastropods are long and thin and often the same color as the gorgonian on which they are living. Apparently they are able to incorporate the pigments of the gorgonian into their own tissue. They thus become camouflaged and less visible to potential predators.

One gastropod which is often found near the base of gorgonian colonies and on reef corals is the coral snail *(Coralliophila* spp.). The coral snail appears to be more of a parasite of the gorgonians than a predator. It lacks a radula which is used in most snails for cutting into shells or grinding down food. Instead, the coral snail produces special enzymatic secretions in its salivary gland which allows its long proboscis to penetrate the tissues of the coral polyps. The salivary secretion might also act as an anesthetic to protect the snail from the gorgonian's stinging cells.

A number of predators of gorgonians are so small that they are rarely seen by divers. These include colorful sea slugs *(Tritonia* spp.), minute shell-less snails called aplacophorans, and creeping combjellies (ctenophores). Small pontoniid shrimp, which like the snails of the genus *Neosimnia* have evolved a purple or yellow protective coloration, also live on the branches of gorgonians.

Finally, mention should be made of the various reef fish that occasionally can be seen nibbling away at the branches of gorgonians. Spadefish *(Chaetodipterus faber)*, butterflyfish *(Chaetodon capistratus)*, angelfish *(Pomacanthus arcuatus)*, filefish *(Alutera scripta)*, and cowfish *(Acanthostracion quadricornis)* have all been found with the remains of gorgonian polyps in their stomachs, but it should be noted that the diet of all of these fish consists primarily of other types of marine organisms, and not gorgonians. Considering the great abundance of gorgonians on reef areas of the West Indies, it is surprising that no fishes have been reported to feed exclusively on gorgonian corals. Possibly it is the stinging cells (nematocysts) of gorgonians or some noxious substance released by the polyps that discourages would-be predators. Many of the small invertebrates that crawl over the surface of gorgonian colonies secrete a thick layer of mucus which apparently protects them from the powerful sting of the nematocysts.

NEMATOCYSTS

Nematocysts are one of the features which gorgonians have in common with all other coelenterates. They are responsible for the notorious reputations of stinging coral, hydroids, various jellyfish, and some types of sea anemones. Fortunately, the stinging cells of gorgonians are not very potent and consequently, there have been few ceses of divers being stung by these corals. However, divers should be careful to avoid gorgonian branches which are encrusted with stinging coral. Such branches are light colored and often swollen in appearance.

Nematocysts are actually cell organelles in the sense that each one is only a specialized part of a single cell. Structurally a nematocyst consists of an oval hollow capsule with a thread or filament at one end. There are a dozen or more different types of nematocysts; some with enormous capsules and small bulbous threads, and others with small capsules but extremely long threads. The threads can be smooth surfaced or covered with sharp spines of assorted sizes and shapes. In the most dangerous types, the capsule is fitted with a poisonous fluid which is expelled through the tip of the thread. In an inactive state the thread is inverted and coiled up inside the capsule which is primed to explode like a cannon.

Although it may not seem so from their plant-like appearance and sluggish nature, most coelenterates are highly successful carnivores, and it is with their nematocysts that they catch their prey. When an animal such as a small shrimp brushes against a tentacle of a coelenterate, the nematocysts are discharged and the threads explode out of the capsules. Nematocysts with long spineless threads are used to entangle the shrimp like lassos, while those with spines snare the prey like harpoons. Once captured the shrimp is quickly immobilized by the nematocyst toxin, then transported to the mouth and ingested.

In most coelenterates the toxin is only strong enough to kill small organisms, but in one species of Pacific jellyfish the poison is reported to be more deadly than that of a cobra. This species has been implicated in a number of human fatalities. Fortunately, it does not occur in the Atlantic. But divers and swimmers in Caribbean waters should avoid the Portuguese Man-of-War. This siphonophore (it is not a true jellyfish, but related instead to hydroids and stinging coral) has very long tentacles packed with powerful stinging cells.

A Portuguese Man-of-War, *Physalia physalis*, is, like a gorgonian, a colonial organism made up of numerous small polyps. In a *Physalia* there are three different kinds of polyps each specialized for catching prey, or for feeding, or for reproduction, but in a gorgonian all the polyps are identical and perform all the functions necessary for survival.

THE POLYPS

A gorgonian polyp has a very simple, radially symmetrical body plan. Basically, the polyp is a long cylindrical tube sealed off at one end, but with an oral opening or mouth at the center of the opposite end. Eight tentacles are arranged in a circle around the mouth, and each tentacle is pinnately branched like a feather. The polyps of sea pens, sea pansies and other soft corals also have eight branched tentacles, and because of this similarity, these coelenterates, along with the gorgonians, are classified together in a group called the Octocorallia. The sea anemones and reef corals, however, usually have six or some multiple of six, unbranched tentacles, and for this reason they, with a number of smaller groups, are classified together in the Hexacorallia.

The word polyp is derived from the French word Poulpes which means octopus. The name was first applied to coral animals in the 1740s, possibly because of the superficial resemblance between the tentacles of a polyp and the outstretched arms of an octopus. Some naturalists of that time thought that the polyps were "insects or worms" which inhabited the coral, but later workers proved that the polyps were an integral part of each coral colony.

The interior of a gorgonian polyp is simply a large cavity called a coelenteron which, in its upper part, is subdivided radially into eight pie-shaped sections by partitions of tissue known as septa. The septa stretch across the coelenteron from the inner wall of the polyp to the esophagus. The esophagus, as the name suggests, is a tube which extends down into the coelenteron from the mouth.

Apart from these structures, a gorgonian polyp does not contain the complex organ systems found in higher animals. There are no special organs of digestion, excretion, respiration or circulation. Food is taken in through the mouth, digested in the coelenteron, and undigested material is expelled through the mouth. Oxygen is extracted from the surrounding water, and carbon dioxide given off, through the general surface epithelium of the polyp. The nervous system consists of a simple nerve "net" which allows for only a limited amount of coordinated movement.

The simplicity of the polyp structure is also reflected in the microscopic anatomy of its body tissues. The body wall and tentacles of a polyp are made up of three layers of tissue; an outer epidermis, the inner gastrodermis, and a middle layer of jelly-like material called mesogloea. A thick layer of mesogloea is responsible for the gelatin-like consistency of the common jellyfish, but in gorgonians this layer of tissue is usually relatively thin. While the mesogloea rarely contains cells, and the gastrodermis is made up of mostly digestive gland cells, the epidermis contains a large assortment of different cells, including mucus producing cells and ciliated cells for protecting the outer surface of the polyp by removing particles of silt and sand, plus contractile cells for movement, sensory cells, and also the specialized nematocysts which are highly modified for catching food.

Although there is no special circulatory system in the polyps of gorgonians, the cells on the two sides of the esophagus are provided with long flagellae whose coordinated movements cause a current of water to flow into and out of the polyp. Also all the polyps of a colony are united, in a limited way, by means of narrow canals which meander through the tissue between the polyps. This tissue is called coenenchyme, and it usually contains the microscopic calcareous structures already referred to as spicules. Spicules come in diverse sizes, shapes and colors, and they are a key character in classifying and identifying different species of gorgonians. In some species spicules are also found in the body wall of the polyps and also at the base of the tentacles.

REPRODUCTION

In gorgonians the reproductive system is not very complex. The eggs and sperm develop in restricted areas of tissue on the septa. All of the polyps of one colony are apparently of the same sex. In most species fertilization takes place within the polyps containing the eggs. The embryos develop into flattened, oval-shaped masses of cells called planulae. Planulae adopt a free-swimming planktonic existence which lasts a variable length of time. They drift with the currents and are thus scattered over great distances. During this time they are susceptible to heavy predation from other planktonic organisms and from benthic animals which feed on plankton. An excess number of planulae are released from each gorgonian colony to insure the survival of at least a few. One recent study has shown that an average size colony of one species is capable of producing hundreds of thousands of planulae.

During the reproductive season, some gorgonians can be seen to be covered and surrounded by clouds of tiny worm-like planulae, each of which is a fraction of the size of a grain of rice. In these species the eggs are retained in the polyps until they grow into fully developed planulae, and then they are released simultaneously.

Eventually, the planulae give up their planktonic life, settle to the bottom and attach themselves to any available hard substrate. As a tentacle buds and a mouth appears at the free end of the newly settled larvae, the planulae become transformed into polyps. An axial skeleton is then secreted by a layer of cells at the attached end of the polyp, and calcareous spicules begin to appear in the surrounding tissues. As the skeleton increases in size additional polyps develop asexually at the upper end of the colony. Eventually the colony may reach a large enough size so as to contain more than 50,000 polyps.

A gorgonian colony grows at the rate of several inches per year, but newly settled colonies might grow faster. Without a higher growth rate smaller colonies would have a greater chance of being smothered by silt and sediment or of being overgrown by neighboring marine organisms. The rate of upward growth probably slows down as a gorgonian reaches its maximum size. An average size colony, about 2 ft. high, may be anywhere from 10-15 years old and the biggest colonies may be several times older. Thus a stand of gorgonians is a relatively permanent fixture on a coral reef and only through the ravages of a natural disaster such as a hurricane, or the carelessness of man, is the ecosystem upset enough to destroy in a short time what it has taken nature many years to build.

☆

WEST INDIAN REEF MOLLUSKS

ROBERT C. WORK
Research Scientist, School of Marine and Atmospheric Science, University of Miami

Although some West Indian shells reached Europe much earlier, it was not until the late seventeenth and early eighteenth centuries that a fair variety of species began appearing in the cabinets of the rapidly growing number of curio collectors. By the end of the eighteenth century, the pink conch graced countless mantels and parlor curio cabinets; and many less common shells such as the mouse cowry, crown cone, and music volute had found their way into the private collections of ardent hobbyists throughout the Old World. Some of the rare shells of those days are now considered quite common, while others still remain choice collectors' items. During the past few centuries many West Indian reef shells were known only from beach specimens or from fish traps; and although some reef-dwelling shells were obtained by wading, most species did not become readily available to collectors until the popularization of the diving mask and swim flippers, and later SCUBA, after World War II. Today shell collecting is a very popular hobby, and many collectors spend their holidays diving on West Indian reefs in quest of specimens.

Five of the six classes of the phylum Mollusca may be encountered in areas of reef; however, the class Scaphopoda, the tooth or tusk shells, is represented in the reef habitat only by occasional specimens in pockets of sand. The class Gastropoda, comprised of the numerous limpets, conchs, snails, and sea slugs, is well represented on the reefs, as is the class Pelecypoda (Bivalvia), the various oysters, clams, and scallops. The class Cephalopoda consists of the octopuses, squids, the squid-like spirula, several species of paper nautilus, and several species of chambered nautilus. The latter are found only in the southwest Pacific. On the West Indian reefs the class is represented only by the squids and octopuses, although emergent sandbars and outcroppings in reefy areas may yield the stranded shells of the spirula and paper nautiluses. Some of the girdled, eight-valved chitons, class Amphineura, are also found on the reef, but they are not as conspicuous as those found on rocky shores in the Caribbean.

The sixth class, Monoplacophora, is represented by only a few species of deep-sea mollusks, which are sometimes referred to as gastroverms. These unusual animals, known only as fossils until discovered alive in 1957, possess a single limpet-like shell covering a segmented body, each segment of which has a duplicate set of the vital internal organs.

During the daylight hours, few mollusks are immediately evident on the Caribbean reefs. A diver may see a fleeting school of squid, a pink conch leaping clumsily near the edge of the reef, a ruffled nudibranch gliding over a sponge, or a flamingo tongue snail clinging to the lavender latticework of a sea fan. Possibly he will see more, perhaps less.

The adventuresome night diver will not be confronted with such a lack of diversity; for in the darkness the octopus scuttles about, while tritons, cone shells, shaggy-mantled cowries, and many other beautiful gastropods venture from the protective hollows and crevices of the reef framework and from beneath slabs of rock and coral heads.

The bivalves of the reef are so secreted or so well camouflaged by encrusting growths that they go largely undetected at all times. The beautiful spiny oysters and leafy rock oysters may be so covered with sponges, algae, and other growths that the only indication of their presence is the sudden snapping shut of the valves as the diver swims by. The file shells and rock scallops live cloistered in crevices or beneath rocks and corals, while date mussels and other boring bivalves are completely hidden within the hardness of limestone boulders and living coral heads.

Although their shells are frequently so encrusted that they go unnoticed, certain reef dwelling gastropods are often completely exposed to the view of the diver. The large carved star shell, *Astraea caelata,* and the many varieties of the imbricated star shell, *Astraea tecta,* are often thus exposed but undetected. On the other hand, flamingo tongues and others species of the genus *Cyphoma* are so easily located on their host gorgonians that they have been greatly over-collected in some areas. The large pink conch, *Strombus gigas,* and the milk conch, *Strombus costatus,* are also often seen as they move about in the open. The hawk-wing conch, *Strombus raninus,* and the quite rare rooster conch, *Strombus gallus,* are usually not concealed, although they may sometimes be partially buried or hidden under the over-hanging edges of rocks and corals. These species of *Strombus* are not typically reef animals, but they do occur around patch reefs and on the shoreward side of fringing reefs where they graze on algae on the rubble strewn bottom or in patches of turtle grass. Reports of *Strombus* being carnivorous are totally unfounded. The flaring lip of the shell of *Strombus* is developed only at maturity, and young specimens, commonly called "rollers," are more cone-shaped.

The helmet shells may also be found in the rubble areas of the reef. However, the very large emperor helmet, *Cassis madagascariensis,* is much less likely to wander reefward from a sandy habitat than are the king helmet, *Cassis tuberosa,* and the flame helmet, *Cassis flammea.* The small cowry-helmet, *Cypraecassis testiculus,* is very common on some of the Caribbean reefs. The related bonnet shells, *Phalium granulatum* and *Phalium cicatricosum* are found only in sandy areas, but they are sometimes encountered near the edge of the reef. *Phalium* will eat only certain types of heart urchins, sand dollars, and sea biscuits, while *Cassis* and *Cypraecassis* will eat a much wider variety of echinoids including the typical sea urchins. The latter are completely ignored by *Phalium.*

Another large gastropod of the reef is the trumpet triton, *Charonia variegata*; but, as with other large gastropods of the Caribbean reefs, *Charonia* is not restricted to the reef environment. This handsome snail hides by day in caves or under coral heads or ledges. The trumpet triton's food consists of starfish, sea urchins, and sea cucumbers, but on the Caribbean reefs its diet is largely starfish-free. Starfish are infrequent on these reefs, although species of *Ophidiaster, Linckia* and *Asterina* do occur on the reefs proper. Other species of starfish, relished by the trumpet triton, do occur in considerable numbers in other Caribbean habitats. The partridge tun shell, *Tonna maculosa,* sometimes found in areas of reef, shares the trumpet triton's fondness for sea cucumbers.

Closely related to the trumpet triton are a number of other Caribbean tritons of the genus *Cymatium,* the shells of which are usually covered with a hair-like periostracum and the animals of which are usually beautifully spotted or mottled, as are also the animals of the genus *Charonia.* Although the feeding habits of all Caribbean *Cymatium* have not been determined, certain species are known to feed on other mollusks. Female tritons brood their eggs, remaining on the egg masses until all of the free-swimming larvae hatch. Due to a long, pelagic larval life, some species have been carried along by ocean currents to become circumtropical in distribution. Several species of frog shells, *Bursa,* are found on the West Indian reefs, and although they closely resemble the tritons, they belong to another family.

Among the most interesting reef dwelling mollusks are the cone shells, which are the favorites of many shell collectors. The cones are particularly fascinating because of their method of capturing prey. Their hollow, harpoon-like radular teeth are connected to a poison secreting organ and are used for stunning or killing prey, which is then swallowed whole. They strike their victims with their extensible proboscis in a manner reminiscent of a striking snake. Although the food of many species is unknown, some Pacific species feed on worms, others on gastropods, and others on fish. The food of only a few Caribbean species is known. Many species have much smaller radular teeth and possibly less potent venom than do certain others such as the fish eaters and some of the gastropod eaters. Some Pacific cones even attack and eat other cones, and at least one species is exclusively a cone eater. At any rate, all cones should be handled with care, for all are venomous

continued on page 116

Descriptions of the following pictures

page 97

UNUSUAL GROWTH OF ELKHORN CORAL

An unusual growth form of elkhorn coral (Acropora palmata). *In this form, the branches continue to taper distally until the smallest diameter of each branch is reached at the tip. Typically the branches of this species flare out in a palmate fashion as seen in the following photograph. These picturesque forms of elkhorn coral are found especially around the Caicos Islands.*

page 98 (upper)

ELKHORN CORAL

Normal growth of elkhorn coral as found throughout the West Indies and the Florida Keys.

page 98 (lower)

STAGHORN CORAL

French grunts (Haemulon flavolineatum) *swim above a dense entanglement of staghorn coral* (Acropora cervicornis).

page 99

LIVE POROUS CORAL AND CHRISTMAS TREE WORM

A colony of porous coral (Porites porites) *inhabited by serpulid worms, commonly called plume worms or, in this form, Christmas tree worms. Serpulids construct solid, calcareous tubes, which are often overgrown by coral, leaving only the aperture of the tube exposed. In some species the terminus of the tube is armed with a sharp spine, which can give a serious cut to a diver. Clearly visible in the photograph is the pink, antler-like process on the operculum of the specimen in the center.*

page 100 (upper)

CLOSE-UP OF BRAIN CORAL

Here is a living brain coral (Diploria labyrinthiformis) *with tentacles extended for feeding. In brain corals, as the coral grows, the original polyp widens and adds more tentacles, then elongates and develops a series of mouths along the coral's meandering furrows, which are lined on each side by a single row of tentacles.*

page 100 (lower)

CLOSE-UP OF STAR CORAL

This close-up view of living star coral (Montastrea cavernosa) *shows clearly the expanded polyps. In the star corals, the polyp secretes a wall across the middle of its cup and thus divides to become two polyps. As this process is repeated again and again, a colony of polyps is formed, all sharing a common skeleton of limestone.*

page 101 (upper)

ELKHORN CORAL

Standing like sentinels against a brooding underwater sky, these unusual colonies of elkhorn coral (Acropora palmata) *resemble arborescent cactus plants. This growth habit of these specimens is very similar to that of pillar coral* (Dendrogyra cylindrus). *See also page 97.*

page 101 (lower)

EXTENDED POLYPS OF STOKES' STAR CORAL

A beautiful specimen of Stokes' star coral (Dichocoenia stokesii) *with extended polyps is shown in this picture. The small, green rosette near the coral is the calcareous alga* Rhipocephalus phoenix.

page 102 (upper)

FIRE CORAL

Like a multi-turreted castle, this growth of fire coral (Millepora *sp.*) *stands out against the dark water. It is not a true coral but a hydrozoan, with stings that are well-known and feared by divers.*

page 102 (lower)

BLUE TANG AND FIRE CORAL

Blue tang (Acanthurus coeruleus) *swim near a large colony of fire coral* (Millepora *sp.*). *The sea fan in the foreground has been completely encrusted by fire coral. Fire coral often encrusts other objects, and divers sometimes find old bottles which are completely encrusted with beautiful formations of fire coral.*

continued on page 113

UNUSUAL GROWTH OF ELKHORN CORAL

ELKHORN CORAL

STAGHORN CORAL

LIVE POROUS CORAL AND CHRISTMAS TREE WORM

CLOSE-UP OF BRAIN CORAL

CLOSE-UP OF STAR CORAL

COLONY OF ELKHORN CORAL

EXTENDED POLYPS OF STOKE'S STAR CORAL

FIRE CORAL

BLUE TANG, SEA FAN AND FIRE CORAL

*LAVENDER SEA FEATHERS,
CORAL AND SPONGES*

CLOSE-UP OF A GORGONIAN

GORGONIANS FOR PHARMACEUTICAL USE

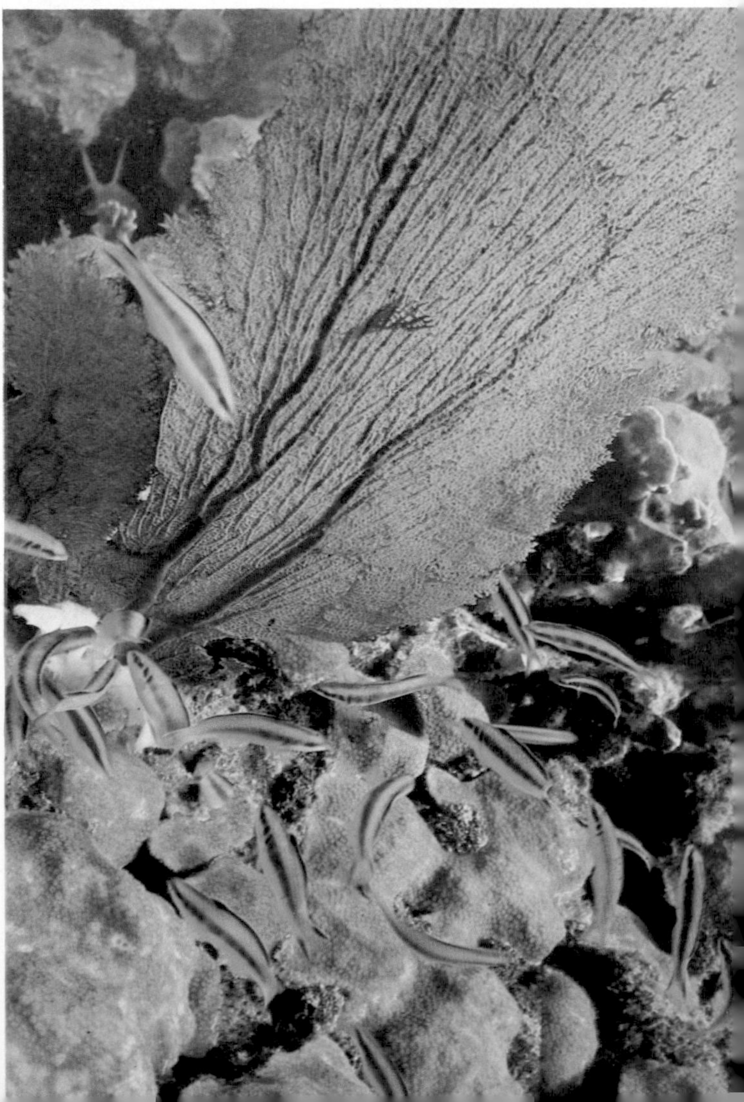

SEA FAN WITH SCHOOL OF BLUE HEAD WRASSE

FAN-SHAPED GORGONIANS, TWO HUNDRED FEET BELOW SURFACE

REEF LIFE:
STOPLIGHT PARROTFISH,
BRAIN CORAL, FIRE CORAL,
AND
PLEXAURID GORGONIANS

TUNICATES AND HYDROIDS

SEA ANEMONE

GIANT SEA ANEMONE

CARIBBEAN WARTY ANEMONE

PINK FEATHERDUSTER WORMS

SERPULID OR CHRISTMAS TREE WORM

BRISTLEWORM

SABELLID OR FAN WORM

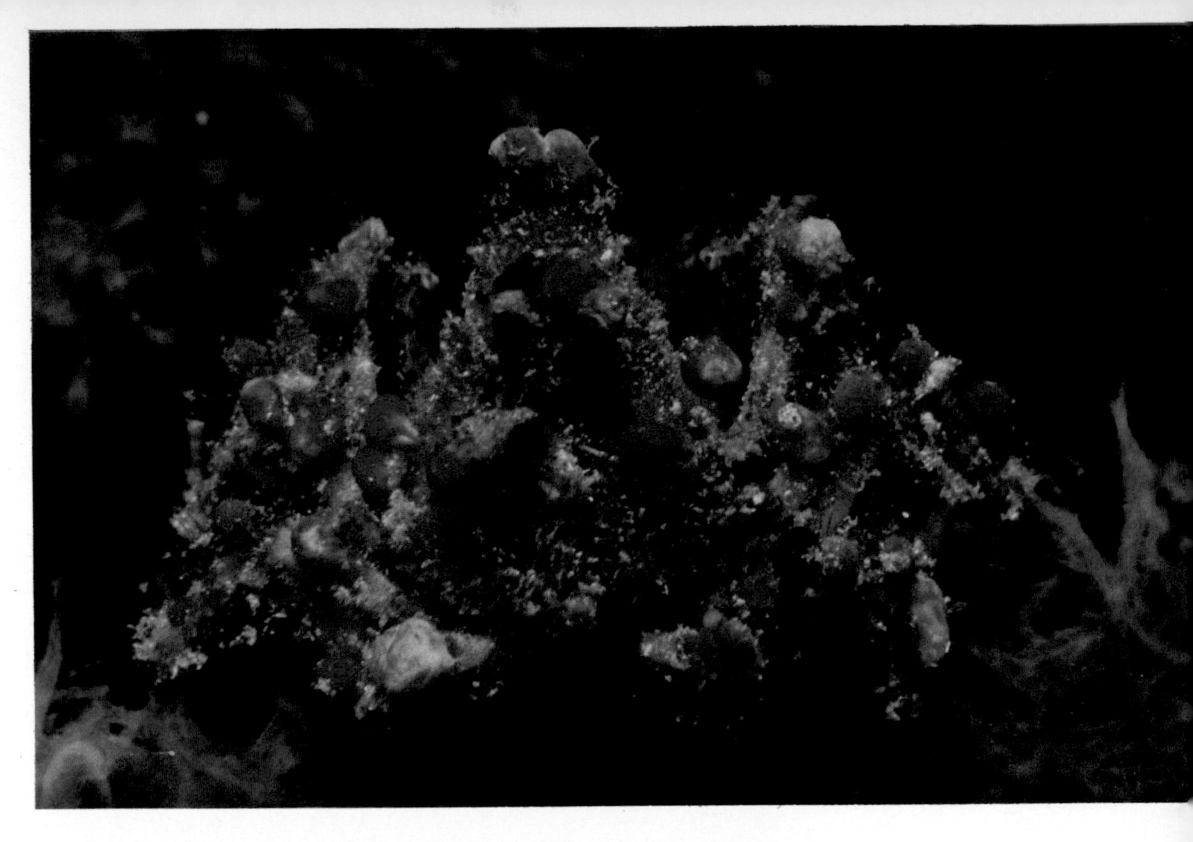

SPIDER CRAB COVERED WITH SPONGES, ALGAE AND ASCIDIANS

SPINY LOBSTERS MARCHING BEFORE SARGASSUM

SHELLS OF THE SPOTTED FLAMINGO TONGUE SNAIL

← *SPOTTED FLAMINGO TONGUE SNAIL ON SEA FAN*

NUDIBRANCH

RUFFLED SEA SLUG

PINK CONCH

KING HELMET SHELL

ROYAL FLORIDA MITER

ATLANTIC TRUMPET TRITON

AMBER PEN SHELL

Descriptions of the foregoing pictures

(Continued from page 96)

page 103 (upper)

LAVENDER SEA FEATHERS, CORAL AND SPONGES

A symphony of colors and forms, it is a typical community of the coral reef. Lavender sea feathers (Pseudopterogorgia bipinnata) *grow on a rocky ledge along with corals and sponges. The yellowish-green branching sponge is a species of* Verongia.

page 103 (lower)

CLOSE-UP OF A GORGONIAN

Frosty, eight-tentacled polyps create a beautifully intricate pattern along the branches of a gorgonian (Iciligorgia schrammi). *This striking gorgonian grows only on the deeper parts of the reef.*

page 104 (upper)

GORGONIANS FOR PHARMACEUTICAL USE

Many gorgonians have been tested for medicinal properties. The robust specimen in the foreground is Plexaurella dichotoma. *The dark specimen, with a main axis that is more or less horizontal in the photograph, is* Plexaura homomalla, *currently the most important gorgonian to the pharmaceutical industry because it produces prostaglandins, basis for the "once-a-month" birth control pill. Just what a gorgonian does with a mammalian hormone-stimulating chemical is still a mystery.* Plexaura homomalla *is abundant.*

page 104 (lower)

SEA FAN WITH SCHOOL OF BLUEHEAD WRASSE

A school of bluehead wrasse (Thalassoma bifasciatum) *gather at the base of a sea fan* (Gorgonia ventalina). *The juveniles, mature females, and mature normal males all display the colors shown here. Only the super-male* (not seen here), *with its beautiful blue head, rightfully lives up to its common name. The massive coral is* Montastrea annularis.

page 105 (upper)

FAN-SHAPED GORGONIANS, TWO HUNDRED FEET BELOW THE SURFACE

The colors of gorgonians growing in deeper parts of the reef or ocean floor are often pinkish or flesh colored. These fan-shaped specimens (Iciligorgia schrammi) *grow on a reef at two hundred feet depth.*

page 105 (lower)

REEF LIFE

A stoplight parrotfish (Sparisoma viride) *swims on a reef composed of fire coral* (Millepora *sp.*), *brain coral* (Diploria strigosa), *and plexaurid gorgonians. The coloration of this parrotfish indicates that it is a terminal-phase male. The young and intermediate phases are colored quite differently.*

page 106

WONDER OF THE REEF: TUNICATES AND HYDROIDS

A cluster of colonial ascidians or tunicates, commonly called sea squirts, grows on the branches of a gorgonian. The apertures of both the incurrent and excurrent siphons are fully open in all of the zooids comprising this colony. The delicate, plume-like structures are hydroids with some polyps fully expanded. Many polyps of the gorgonian are also expanded.

page 107 (upper left)

SEA ANEMONE

Two striking examples of sea anemone (Calliactis tricolor). *This particular species is sometimes called the hermit crab anemone due to its frequent occurrence on shells occupied by hermit crabs. This species is reported from North Carolina and the Gulf of Mexico southward to Brazil.*

page 107 (upper right)

GIANT SEA ANEMONE

Two anemones (Condylactis gigantea) *nestle peacefully among sponges and coral, where their stinging and poisonous tentacles are a real death trap for plankton and other small animals. These tentacles are also a valuable defense against fish and other larger animals, which avoid touching these stinging arms.*

page 107 (lower)
CARIBBEAN WARTY ANEMONE
It seems unfair that such a beautiful animal is commonly called the Caribbean warty anemone (Bunodosoma granuliferum). *Several other species of* Bunodosoma, *especially* Bunodosoma cavernatum *of the southeastern U.S., closely resemble this species.*

page 108 (upper left)
SERPULID OR CHRISTMAS TREE WORM
The beautiful spiral gills of a serpulid worm. Compare these structures with those of the sabellid worms (opposite above and below, same page).

page 108 (upper right)
PINK FEATHERDUSTER WORMS
Featherdusters or sabellid worms spread here their tentacular crowns, resembling a bouquet of pink flowers. The soft, non-calcareous tubes of the sabellids are clearly visible in the picture.

page 108 (lower left)
BRISTLEWORM
In spite of their appearance, the plume worms and feather duster worms are polychaetous annelid worms closely related to this amphinomid polychaete or bristleworm (Hermodice carunculata), *seen here devouring the polyps of a coral* (Tubastraea tenuilamellosa). *The calices of the coral to the right of the worm appear already to have been cleaned of living tissue. Amphinomid polychaetes, often called fire worms or bristleworms, should not be touched; for the bristles or setae, seen here as white tufts, cause great pain upon penetrating and breaking off in one's flesh.*

page 108 (lower right)
CROWN OF FEATHERDUSTER OR FAN WORM
The magnificent tentacular crown of a sabellid worm. The sabellids are commonly called featherdusters or fan worms, the former name also being applied to serpulids. Sabellids always build soft tubes that are not calcified and therefore can readily be distinguished from the serpulids, which always have calcareous tubes. Sabellids do not have opercula as do most serpulids. In both families the feather-like structures are gills used for respiration and filter feeding.

page 109 (upper)
SPIDER CRAB COVERED WITH SPONGES, ALGAE AND ASCIDIANS
That's camouflage. This species of spider crab (family Majidae) has covered itself with bits of algae, sponges, and ascidians. This species and related species with similar habits are often called decorator crabs. The tiny red eyes are scarcely noticeable between the spotted, purple chelipeds (pincer bearing legs), which are free of disguising growths.

page 109 (lower)
SPINY LOBSTERS BEFORE SARGASSUM
Spiny lobsters (Panulirus argus) *march before a forest of* Sargassum *during one of their mysterious migrations. With some frequency divers report long columns of migrating spiny lobsters; but, as yet, no one knows the reason for such mass movements.*

page 110 (upper left)
SPOTTED FLAMINGO TONGUE SNAIL ON SEA FAN
A spotted mantle covers the shell of a flamingo tongue snail (Cyphoma gibbosum) *as it feeds on the tissues of a yellow sea fan* (Gorgonia flabellum). *There are three species of* Gorgonia *or true sea fans in the West Indian region, and two of these species have more than one form.* Gorgonia flabellum *is very rare in Florida and appears to be absent from Bermuda.* Gorgonia ventalina *is the common sea fan of Bermuda and Florida, but it also occurs commonly in the West Indies.*

page 110 (upper right)
SHELLS OF FLAMINGO TONGUE SNAILS
A collection of flamingo tongue shells. The specimen at the extreme lower right and that at the left (vertically positioned) are shells of Cyphoma mcgintyi. *The remainder are* Cyphoma gibbosum, *shown alive in the opposite photograph. The former appears to be confined to the southeastern United States, the Gulf of Mexico, and the Bahamas.* Cyphoma gibbosum *has a more general West Indian distribution.*

page 110 (lower left)
NUDIBRANCH
This sea slug or nudibranch (Chromodoris neona) *is all dressed up in an incredible pattern of red, white, and blue. The pair of purple rhinophores (sensory organs) and the purple-edged gills are retractile and completely disappear from view if disturbed. This beautiful species occurs from Florida southward to Brazil.*

page 110 (lower right)
RUFFLED SEA SLUG
This ruffled sea slug (Tridachia crispata) *displays a remarkable combination of translucent pastel shades. The colors of this species are quite variable, but the beauty of this particular specimen is rarely surpassed. This species is known from Bermuda, Florida, and many Caribbean localities.*

page 111 (upper left)
PINK CONCH
The pink conch (Strombus gigas), *in addition to being the number one souvenir shell of the Caribbean, Florida and Bahamas, is an important food to many of the island people. The specimen shown here has not yet developed the flaring lip which characterizes the mature shell.*

page 111 (upper right)
KING HELMET SHELL
The apertural view of the handsome king helmet shell (Cassis tuberosa). *Shells of* Cassis *are highly prized for the beautiful polish on the flaring parietal shield and outer lip. Cameos are cut from the shells of certain species of* Cassis.

page 111 (lower left)
ROYAL FLORIDA MITER
The beautiful and rare royal Florida miter (Mitra florida). *This specimen, collected off the Florida Keys, is one of the few known to have been collected alive. The distribution of this species is very poorly known.*

page 111 (lower right)
ATLANTIC TRUMPET TRITON
Both the animal and shell of the Atlantic trumpet triton (Charonia variegata) *are handsomely colored. A tiny black eye can be seen at the base of the tentacle in the foreground. At the rear of the foot is the dark operculum, which seals the shell's aperture and protects the animal when he withdraws.*

page 112
AMBER PEN SHELL
The translucent beauty of the valves of an amber pen shell (Pinna carnea) *is displayed on driftwood against a sun-dappled West Indian sea. The beautiful "Cloth of Gold" is woven from the byssal threads (anchoring filaments) secreted by* Pinna, *common all over the West Indies, Florida Keys and Bermuda.*

☆

(Continued from page 95)

to some degree. Some of the larger Pacific species, including known fish eaters and gastropod eaters, have been responsible for agonizing human deaths.

Although a number of reef-dwelling cones may be found in the Caribbean region, only two of these may be considered common over much of the area. These are the crown cone, *Conus regius,* and the mouse cone, *Conus mus.* The crown cone reaches a fairly large size and occurs in an extraordinarily wide range of colors and color patterns. The smaller and much less variable mouse cone has brownish mottlings on a blue-grey background. The crown cone and mouse cone feed on polychaete worms of the family Amphinomidae.

Other species of general Caribbean distribution that may be encountered in areas of reef are the yellow to orange-red carrot cone, *Conus daucus;* the small cardinal cone, *Conus cardinalis,* with brown and white markings on red; and the exceptionally beautiful granulated cone, *Conus granulatus.* This latter species, sometimes called "Glory of the Atlantic," occurs in shades of rose and bright red often covered with flecks of brown; and although on some Caribbean islands worn beach specimens are not rare, living specimens are exceedingly difficult to obtain. Some of the largest and most beautiful specimens known have come from shallow reefs in the Florida Keys.

Certain of the showy Caribbean reef cones are far more likely to be found in the lower Caribbean, especially on the reefs of Curaçao, Bonaire, and Aruba. Among these are the hieroglyphic cone, *Conus hieroglyphus;* the antillean cone, *Conus insularis;* the golden cone, *Conus aurantius;* and the Atlantic agate cone, *Conus ermineus.* This latter species has been wrongly known as *Conus ranunculus* in the Caribbean and *Conus testudinarius* in West Africa, where it also occurs. In addition *Conus purpurascens* of the Eastern Pacific is believed by some to be the same species as the Atlantic agate cone. If they are not the same, they are certainly very closely related, and therefore divers in the Caribbean should exercise exceptional care in handling living examples of the agate cone; for its Pacific counterpart is a known fish eater and possibly quite dangerous to man. The Atlantic agate cone reaches a fairly large size and has a rather great range of coloration. Specimens may be white with pale yellow or orangish markings, or they may have a bluish or purplish background covered with light to dark brown markings. A helpful identifying feature is the smooth, rounded shoulder of the shell's spire. This species is rare on the Florida reefs, but it has been found alive on the southeast coast as far north as Palm Beach County.

On Florida reefs the enormous horse conch, *Pleuroploca gigantea,* is sometimes found. This species, one of the largest gastropods in the world, has been officially designated the state shell of Florida. It is a voracious predator on other mollusks, even attacking mature pink conchs. The related smaller but equally predaceous tulip shell, *Fasciolaria tulipa,* is also sometimes found in areas of reef, but it is more common in grassy areas. The horse conch does not occur in the Caribbean nor in the Bahamas; however, in the Bahamas, Cuba, Central America, and the southwest Gulf of Mexico there is a large gastropod superficially resembling *Pleuroploca* and often wrongly referred to as the horse conch. This is actually the West Indian chank, *Turbinella angulata.* It may be encountered in sandy areas near reefs, where at least a good part of its diet consists of sipunculid worms.

Close relatives of the West Indian chank are the vase shells. The common vase shell, *Vasum muricatum,* found in Florida and parts of the Caribbean, is not typically a reef dweller, but it may be found in the reef habitat. The spiny vase, *Vasum capitellus,* is a reef species and ranges from Puerto Rico southward through the Lesser Antilles. It is most common in the lower Caribbean, especially around the Netherlands Antilles.

The entire tropical Western Atlantic has a disappointing variety of cowry shells compared to other tropical areas. Only six species are known from the entire region. However, one of these reaches a larger size than does any other cowry in the world. This is the deer cowry, *Cypraea cervus,* of Florida and the Gulf of Mexico. The smaller, somewhat similar measled cowry, *Cypraea zebra,* has a general West Indian distribution. Two of the Caribbean cowries are considered choice collector's items. One of these is the very rare Surinam cowry, *Cypraea surinamensis,* which ranges from Brazil to Florida. Relatively few specimens are known, most of these having come from stomachs of fish caught on deeper reefs. The other is the mouse cowry, *Cypraea mus,* which appears to be limited to a small area of the Caribbean mainland coasts of Colombia and Venezuela. The two remaining species, *Cypraea cinerea* and *Cypraea spurca acicularis,* are quite small; and both are common on most Caribbean reefs. The cowries brood their eggs much in the manner of *Cymatium.*

Most miter shells of the Caribbean reefs are not particularly showy. However, the royal Florida miter, *Mitra florida,* is considered a great find by collectors. The distribution of this species is poorly known, but a number of specimens have been found on the reefs of the Florida Keys.

Two beautiful species of turban shells may be encountered on the Caribbean reefs. The large channeled turban, *Turbo canaliculatus,* is uncommon but may be found living on very shallow reefs. The threaded turban, *Turbo filosus,* is rare; but apparently specimens are more likely to be found on the reefs off Central America than elsewhere.

Of the many beautiful Caribbean murex shells, most are not reef dwellers, although a few species are found on reefs from time to time. Perhaps the most conspicuous species that wanders onto the reefs is the apple murex, *Murex pomum.* Among other interesting gastropods occurring on the reefs are the well known music volute, *Voluta musica;* various top shells, *Calliostoma;* latirus shells, *Latirus;* and coffee-bean shells, *Trivia.* There are, of course, many other reef dwelling gastropods, including a number of limpets and beautifully colored nudibranchs or sea slugs.

There are fewer showy bivalves than gastropods on the reefs, but some of them are exceptionally beautiful beneath their often heavy encrustations. The American spiny oyster, *Spondylus americanus,* occurring in an unusually diverse range of colors, is one of the world's most beautiful shells. It is uncommon in relatively shallow water, and specimens from such depths are rarely attractive. Specimens from the deeper reefs are often remarkably spinose, especially those from the areas of gravel-like rubble where the attached valve of the shell may develop to its fullest. Specimens from solid reef structures or from old wrecks often have the attached valve greatly flattened and relatively spineless. Another spiny oyster, *Spondylus ictericus,* is fairly common on the shallow portions of some reefs. Its color is often brick-red, and it does not attain a size as large nor spines as long as those of the American spiny oyster.

Somewhat resembling the spiny oysters are the leafy rock oysters or jewel boxes, *Chama* and *Pseudochama,* but these genera do not have the distinctive "ball and socket" hinge characteristic of *Spondylus.* Well developed specimens of the leafy jewel box, *Chama macerophylla,* when cleaned, resemble chrysanthemums and occur in as many colors. Some other Caribbean jewel boxes are the bright red and white cherry jewel box, *Chama sarda;* the reef jewel box, *Chama sinuosa,* which is one of the most beautiful when the foliations are well developed; and the Florida jewel box, *Chama florida,* which is white with red or pink markings. The left-handed jewel box, *Pseudochama radians,* attaches to the reef by its right valve, as supposedly do all species of *Pseudochama.* Species of *Chama* are believed to always attach by the left valve. However, it has recently been discovered that the rare Caribbean jewel box of Inez, *Pseudochama inezae,* attaches by either valve, a situation that creates quite a problem in molluscan systematics.

There are several species of file shells, *Lima,* in the Caribbean. Although they attach by byssal threads, they are capable of releasing their hold and swimming rather aimlessly for short distances. They are most notable for the beautiful fringe of tentacles bordering the edge of each valve. Depending upon the species, the fringe may be white, lavender, or flaming orange-red.

The rock scallops or rock pectens, *Chlamys,* also attach by byssal threads; and they, too, may swim erratically for brief periods. Some of these are collectors' favorites, particularly the very rare Mildred's scallop, *Chlamys mildredae,* and the little knobby scallop, *Chlamys imbricata.* The maroon-red and white ornate scallop, *Chlamys ornata,* is rare in Florida but common in the Caribbean, while the thorn scallop, *Chlamys sentis,* appears to be more common in Florida. The latter occurs in many colors from white though all shades of orange, purple, and brown. Perhaps the rarest is the equally variably colored *Chlamys multisquamata* from the deeper parts of the reef.

Although from time to time the reef squid, *Sepioteuthis sepiodea,* is seen swimming over the reef, the most frequently encountered cephalopods are the octopuses. Some of the species found on the Caribbean reefs are *Octopus vulgaris, Octopus briareus,* and *Octopus burryi.* The lair of the octopus is often strewn with shells of dead crabs and mollusks from meals past, and occasionally a shell collector finds a rare shell conveniently collected, cleaned and tossed aside by an octopus. The female octopus broods her eggs, constantly stroking them and circulating water around them to keep them clean and well oxygenated until they hatch.

When searching for living mollusks, all divers, whether shell collectors or just interested observers, should exercise extreme consideration for the physical environment as well as for the conservation of the mollusks themselves. Living corals, gorgonians, and sponges should never be dislodged or mutilated. Loose rocks, which are moved or overturned, should be returned to their original positions; and the upper surfaces of the rocks, as found, must be left uppermost, or great mortality of the animal and algal growths on the rocks will ensue. Special care should also be taken to avoid crushing dislodged animals.

During the period between the death of the mollusk and the shell's usual fate as part of the bottom sediment, many shells offer portable shelter to certain other invertebrates, most notably hermit crabs, and stationary shelter to many other invertebrates and fishes. In fact, certain fishes and invertebrates are known to live inside the shells of mollusks while the original tenant is still in residence. Mollusks are an important source of food for many reef inhabitants. Even spiny lobsters and some crabs are able to break away heavy portions of a shell to gain access to the animal. In turn, as has been pointed out, many mollusks are predators on other animals, including other mollusks, while some graze on the algae of the reef. Thus, in addition to being beautiful and fascinating members of the reef community, they play a very important role in reef ecology.

☆

REEF FISHES

PATRICK L. COLIN
Marine Scientist (Coral Reef Fish Behavior and Ecology),
School of Marine and Atmospheric Science, Miami

INTRODUCTION

For the ardent fish watcher to see as much as possible of the piscine life of a reef, he must explore and probe all the reef's many facets, including the near microscopic. Large fishes such as jacks, groupers, parrotfishes, sharks and rays are usually observed without any effort in searching. If the observer stops searching at this level he has missed some of the most important and interesting aspects of the reef ecosystem. There exists a world of small creatures, fishes, invertebrates, and plants, which have a coral patch or reef as their entire universe; living and dying as they have for more years than man has walked the earth, reproducing through often elaborate manner, being eaten by predators which in this Lilliputian world are as ferocious and deadly as the large predators of the land or sea. Some of the fishes which live on this smaller level are the damselfishes, blennies, gobies, jawfishes, miniature sea basses, and basslets.

SOME GROUPS OF COMMON SMAIL REEF FISHES

The damselfishes of the genus *Eupomacentrus* (family Pomacentridae) which include the beau gregory *(E. leucostictus)*, the cocoa damselfish *(E. variabilis)*, the dusky damselfish *(E. fuscus)* and others, might well be considered the most aggressive fishes in the sea. In spite of being only a few inches long, they are not reluctant to chase large fishes from their territories and many divers have been startled by the attack of a damselfish on an ankle, a bit of hair, or right between the eyes of the faceplate. The jaws of these fishes are too small even to scratch the skin of humans but the surprise of the "victim" is usually enough to make them move on. If these fishes were six feet in length and just as aggressive, swimmers would have to stay out of tropical waters and beaches would have to be posted with "DANGER—DAMSELFISH" signs. Fortunately, the damselfishes are very aggressive in response to their small size and their way of life and would not be nearly as pugnacious if they were larger.

Other damselfishes are not as aggressive as the members of *Eupomacentrus*. The blue chromis *(Chromis cyanea)*, an irridescent blue fish which reaches nearly six inches in length, is a peaceful inhabitant of the reef which spends its time during the day picking small food items from the water passing above the reef. The brown chromis *(Chromis multilineata)* is similar in habits and form, but is a drab olive color.

Blennies belong to a few different families of fishes (Blenniidae, Clinidae, Tripterygiidae) and come in a wide variety of species from the larger hairy blenny *(Labrisomus nuchipinnis)* which reaches eight inches, to the hole dwelling blennies *(Acanthemblemaria* spp. and others) some reaching just over one inch in length. Small, slender blennies of the genera *Emblemaria* and *Emblemariopsis* often dwell on living coral heads and are nearly transparent, hence they are very seldom observed by divers. The diminutive arrow blenny *(Lucayablennius zingaro)* reaches one and one-half inches in length, but stalks and swallows prey one-half its own size with a lightning fast lunge forward.

The gobies (family Gobiidae) are superficially similar to the blennies, but have the pelvic fins formed into a sucking disk and two separate dorsal fins instead of one fin as the blennies generally have. These fishes encompass the common small white fish seen skittering about on the sandy margins of reefs, specialized plankton feeders, sponge dwellers, and parasite removers.

The jawfishes (family Opistognathidae) live in burrows they dig in sandy bottoms and are generally small, dusky, and shy fishes who with their head exposed from the burrow wait for some unwary prey to venture too close. The yellowhead jawfish *(Opistognathus aurifrons)*, however, is quite different. It is a plankton feeder found hovering above its burrow, which it occasionally enters by backing in tailfirst, and it ingests small animals selected from the steady stream of planktonic creatures the currents carry by its burrow. Since it is often three or four feet away from its protective burrow, the slightest disturbance is cause to send it fleeing headfirst into the burrow.

The miniature sea basses are members of the family Serranidae which also includes the considerably larger groupers. Included in this group are the brilliantly striped swiss guard basslet *(Liopropoma rubre)*, the carmabi basslet *(L. carmabi)*, the iridescent hamlets *(Hypoplectrus* spp.*)* found in a wide variety of patterns of yellow, blue, orange, brown and white, and the harlequin bass *(Serranus tigrinus)*. Most of these fishes are predators on fishes and invertebrates.

The true basslets (family Grammidae) include the royal gramma or fairy basslet *(Gramma loreto)*, a striking red and yellow fish found generally around overhanging ledges in shallow water, and the deeper water blackcap basslet *(Gramma melacara)*, more subtly colored with purple and black, but still a handsome fish.

THE COLORATION OF REEF FISHES

The iridescent, contrasting colors and unusual patterns of coral reef fishes are perhaps the first thing most people notice. Why are these fishes colored this way? Like most questions in nature there is more than one possible answer.

Some fishes are colored for concealment against the sort of background. A colorful grouper, such as the rock hind *(Epinephelus adscensionis)*, with small orange brown spots over the entire body and larger dark blotches just below the dorsal fin on a light ground color blends well against a backdrop of reef rock and coral. Scorpionfishes (family Scorpaenidae), while not boldly colored, are nearly indistinguishable from the reef backdrop until they move. Other fishes such as the lizardfishes (*Synodus* spp.) and flounders (*Bothus* spp. and others) have light, speckled patterns and resemble closely the sand on which they usually rest.

In general, however, it is believed that the importance of color and pattern to coral fishes is to make themselves conspicuous, not inconspicuous. One aspect of this has been dealt with by the famous German behavioral scientist Konrad Lorenz. He found that individuals of many of the brightly colored reef fishes (termed "plakatfarbig" or poster colored) dwell in only a restricted area and defend this territory against the intrusion by other fishes. The most vigorous defense of the territory is directed at members of the same species. Lorenz felt that such fishes were distinctly and brightly colored in order that they could recognize the members of their own species easily from a distance and avoid each others' territories. This would produce a relatively even distribution of the fishes on the reef and reduce competition for food between members of the same species. The colors may also make the fishes more conspicuous to predators, but the multitude of crevices, caves and holes on the reef probably provide superb hiding places from predators. Good examples of this sort of coloration are the damselfishes such as the beau gregory and the three spot damselfish.

CLEANING SYMBIOSIS

Other fishes use color to advertise services they have to offer. The slender, two inch long neon goby *(Gobiosoma oceanops)* has two iridescent blue stripes and sits conspicuously on coral heads displaying itself. Larger fishes approach the coral head and the gobies swim out, land on the body of the fish and scurry about picking at anything which resembles an external parasite. The gobies are allowed safe passage into and out of the mouth of the fish and enter the gill chamber when the host fish raises the gill chamber.

Several other species of fishes such as the spanish hogfish *(Bodianus rufus)*, spotfin hogfish *(B. pulchellus)*, and young gray and french angelfishes *(Pomacanthus arcuatus* and *P. paru)* are also cleaners. All of the West Indian cleaners are brightly colored and their contrast with their environment helps to make them more visible to potential customers.

One of the most active cleaning fishes is the yellow color phase of the bluehead wrasse *(Thalassoma bifasciatum)*. Another West Indian fish, the mimic blenny *(Hemiemblemaria simulus)*, closely resembles the shape, color, and pattern of the bluehead wrasses and uses this guise to closely approach fishes wanting their parasites removed. Rather than cleaning parasites from the unfortunate host fish, the mimic blenny rushes forward and takes a bite of fins, scales, and skin from its unwitting victim. Fortunately for the fishes wanting to be cleaned the mimic blennies are far fewer in number than the wrasses, and even though the disguise is good, older fishes probably learn to tell the differences between the two after a few bad encounters.

REPRODUCTION OF REEF FISHES

Reproduction in reef fishes is one of the most interesting aspects of their biology. Fishes either lay eggs or bear live young. Among the few live bearing fishes found on West Indian reefs are some of the brotulids such as the black widow *(Stygnobrotula latebricola)* and some sharks and rays. The egg layers come in two general groups, those which lay their eggs on the bottom or some solid object and those who cast their eggs free into the open water.

The species which lay their eggs on the bottom (demersal eggs) include damselfishes, blennies, gobies, toadfishes, and others. This is one reason why some fishes, such as the damselfishes, vigorously defend a territory. When the eggs are laid, the male damselfish guards them, because many other creatures relish fish eggs and are waiting for an opportunity to attack the spawn. A moment's negligence by the guarding male is usually the end for the eggs.

Some bottom spawners have gotten around the problem of protecting the eggs by having one parent, usually the male, pick them up after spawning and brood them in the mouth for a week or so until hatching. The burrow dwelling jawfishes and the shy cardinalfishes both use this method. Interestingly, papa jawfish will occasionally put the eggs down safely within his burrow and come out to feed. After he has eaten, he scoops the eggs up again and resumes his parental duties.

Fishes that cast their eggs into the water (planktonic eggs) include sea basses, angelfishes, parrotfishes, wrasses, jacks, snappers, grunts, and others. Their eggs are generally much smaller and more numerous than those of the bottom spawners and without parental care each egg has less chance of surviving to maturity, hence more eggs are needed.

Some of these fishes, such as the sea basses, have single individuals both male and female during their lifetimes (hermaphroditic). The fish may be a synchronous hermaphrodite, male and female at the same time, or a successive hermaphrodite, being one sex, then the other, but not at the same time. For example, most groupers start out as females and change into males at a larger size.

Even more bizarre are the wrasses and parrotfishes with two types of males. The first is the normal male which is colored the same as the female and participates in mass spawning with dozens of its kind. The second type of male is the "terminal male" or "supermale." For unknown reasons a small percent of the normal females turn into these males which have colors far different from the "normal" fishes. Among these "supermales" is the bluehead phase of the bluehead wrasse *(Thalassoma bifisciatum)*, while the "normal" bluehead wrasses are the yellow color phase males, females and immature individuals. The supermales behave differently from the others also. They are solitary, often territorial and spawn with only a single female at one time. By injecting a normal female parrotfish with testosterone, the male sex hormone, the female will take on the coloration of the supermale. What the ultimate causes of such a change are, and what advantage, if any, it poses for the species possessing it are under investigation by scientists today. However, at present little is known of the relationships.

ACTIVITY OF FISHES

Reef fishes can be divided into three groups on the basis of the periods when they are most active. The first group is the diurnal fishes, those active during periods of daylight. Included in this group are the wrasses, damselfishes, gobies, blennies, surgeonfishes, parrotfishes, angelfishes, butterflyfishes, jawfishes, goatfishes, and basslets. The second group is the nocturnally active fishes. These include cardinalfishes, squirrelfishes, sweepers, and bigeyes. The third group is crepuscularly active. Crepuscle means twilight, and crepuscular refers to fishes most active during the twilight periods of dawn and dusk. These fishes are larger predators such as the groupers.

The changes between night and day on the reef are most interesting. Gone from the reef at night are the groups of colorful fishes so commonly seen during the day. The reef seems deserted at night, but closer examination reveals an abundance of normally hidden life. Pale silvery cardinalfishes hover up to several feet above the reef waiting for organisms in the plankton to come near and be snatched up as morsels of food. Moray eels forage about the bottom using their sensitive olfactory senses to hunt for food. Squirrelfishes and soldierfishes are active predators on invertebrates and fishes. Snappers and grunts, which cluster near bottom cover during the day, forage widely in the area around the reef. Parrotfishes sleep in crevices of the reef with eyes wide open, since they lack eyelids. Many of the smaller parrotfishes are found in mucous covers they secrete possibly to help protect them from predators. The smaller wrasses bury themselves in the sand at dusk and are not seen until the next morning. Round stingrays sit quietly with the front portion of their body arched, possibly to form a "cave" that some unwary creature may enter to its doom.

Ichthyologists have long been engaged in the tasks of collecting, cataloguing, and describing the fishes of the West Indian reefs, a task which still goes on today. The exact number of species of fishes in the West Indies is not known. For a few places, detailed information is available. For example, Dr. James E. Böhlke and associates of the Philadelphia Academy of Natural Sciences have collected over 496 species of fish in the Bahamas, and scientists from the University of Miami have collected some 517 species of fishes in the Florida Keys. Some fishes have limited ranges, hence close examination of the species present at one location in the West Indies does not serve to elucidate the fish fauna of the entire region.

ASSOCIATIONS OF REEF FISHES

Fishes like all organisms are tied closely to their environment. Evolution has adapted them physically to fit into a particular niche in the total ecosystem of the reef. This has resulted in various species of fishes being specialized in regard to factors such as what they eat, how they obtain it, how they reproduce, and how they avoid predation.

Some rather unusual relationships among animals are found on reefs. There is the cleaning behavior mentioned previously. Some fishes are found exclusively associated with sponges. The strange gobies of the genus *Evermannichthys* are only about an inch long, but very slender in relation to their length. They are adapted for moving about the tunnels and chambers of large loggerhead sponges and evidently spend their entire life in one sponge except for a planktonic larval stage. The goby *Risor ruber* has canine teeth on its jaws which point away from, rather than towards, each other. It is also a sponge dweller and must somehow use this strange dentition in moving about or capturing food in the sponge.

Some other gobies are found associated with sea urchins in shallow water. The ninelined goby *(Ginsburgellus novemlineatus)* lives beneath the sea urchin *Echinometra* in water only a few feet deep and feeds upon the tube feet of the urchin above him. A clingfish *(Arcos rubiginosus)* behaves in a similar manner, but a second goby *(Gobiosoma multifasciatum)* also lives under the urchin and evidently does not eat the tube feet. Certainly it is good protection to live beneath a ball of spines like a sea urchin though.

VENOMOUS FISHES AND FISH-POISONING

Although these two terms may sound identical, they can represent two entirely different phenomena. Venomous fishes are those which possess fin spines or other structures modified into a mechanism for injection of venom. Fortunately there are far fewer species of venomous fishes in the West Indies than in the great Indo-Pacific Ocean and those that occur here are generally of less danger to humans than those of the Indo-Pacific.

The stonefish of the Pacific and Indian Oceans is almost legendary in its ability to cause fatalities.

The Atlantic Ocean also lacks the venomous rabbitfishes (Siganidae) and turkeyfishes of the genus *Pterois*. West Indian venomous fishes, however, include scorpionfishes (Scorpaenidae), toadfishes (Batrachoididae), stargazers (Uranoscopidae), eagle rays (Myliobatidae) and other rays (Dasyatidae). Fortunately very few encounters with any of these West Indian fishes are fatal.

A more important problem in the West Indies is fish poisoning caused by the ingestion of certain species of fishes and not related to the possession of any venom apparatus. There are three different types of fish poisoning: puffer-fish, clupeoid, and ciguatoxic. Puffer-fish poisoning, caused by ingesting portions of the internal organs or contaminating the flesh with them, seldom occurs since these fishes are not highly esteemed as food. Clupeoid poisoning has been caused in the Caribbean by herrings (Clupeidae), anchovies (Engraulidae), bonefishes (Albulidae) and tarpons (Elopidae). Fortunately, these species are seldom poisonous and cases of clupeoid poisoning are rare. Little is known of the mechanisms of this type of fish poisoning.

Ciguatera poisoning is the most common type in the West Indies and the list of ciguatoxic fishes includes dozens of species. Toxic fishes occur only in limited areas and a species poisonous in one area may be perfectly "safe" on an adjacent island or bank. Ciguatoxic fishes are most common in the eastern Caribbean, particularly around the Virgin Islands and the northern Lesser Antilles. Generally, poisonous individuals are from water less than 50 fathoms in depth, but this is not an absolute rule.

The most common ciguatoxic fishes are the great barracuda, amberjack, some jacks, some snappers, and some groupers. These are carnivorous fishes feeding on reef or reef related habitats and although no definite mechanism accounting for ciguatera is known, the toxins are evidently produced by some organism in the reef food web and are passed and concentrated in the food chain.

The symptoms of ciguatera poisoning include nausea, vomiting, cramps, tingling of lips and mouth, visual disturbances and on to worse symptoms occasionally resulting in death. If a fish is suspected of being ciguatoxic, it is advisable to feed some of it to a test animal, such as a cat or dog, which can be watched for several hours for symptoms. Be certain that the animal does not regurgitate the fish unnoticed also. If unsure of the situation regarding ciguatera in an area, the local fishermen are usually aware of the local conditions and can be consulted.

HISTORY AND RELATIONSHIPS

Today the greatest variety of marine shore fishes is found near Indonesia and it is believed this is the area from which radiated the fishes which we find today throughout the tropical shore waters of the world. The fishes of this region may have at one time been nearly worldwide in their distribution. During the early Cenozoic (about 50 million years ago) open connections existed between all the tropical oceans with the resultant circum-globe "ocean" being called the Tethys Sea after Tethys, the daughter of Oceanus in Greek mythology. Studies of the fossil fishes found at Monte Bolca, Italy from the Eocene period (about 53 million years before present) show that in many cases the fishes were similar to those found today in the Indo-West Pacific ocean region. During the Miocene period (25 million years ago) the connections of the Mediterranean Sea and the Atlantic Ocean with the Indian Ocean were cut off. Later during the Pliocene period (or early Pleistocene, some 5-3 million years ago) the seaway between the Pacific Ocean and the Caribbean Sea was closed off by the uplift of the Central American land mass.

Although there are only a few species in common today, the fish fauna of the West Indies is similar to the Indo-West Pacific in that many of the families of fishes are found in both oceans. This is the common inheritance of the Tethys Sea. There are, however, dozens of fish families found in Indo-West Pacific which are absent from the Atlantic Ocean. Conversely, there are very few families of Atlantic fishes which are not found in the Indo-West Pacific region. The reason for the difference is believed to be the effects of glaciation.

During the Pleistocene period (1-2 million years before present) there were an estimated 3 to 20 glacial periods when large ice sheets extended from the poles and covered sizeable portions of the continents. This additional ice was produced from the waters of the ocean which lowered sea level perhaps as much as 450 feet. The surface water temperature in the Caribbean Sea was as much as 6°C. (10.5°F.) lower than today and conditions were not particularly benevolent for the survival of tropical marine shore fishes. The Indo-West Pacific region was not affected nearly as much due to its much larger size and relative isolation from the continental land masses. Various families and groups of fishes were eliminated in the Atlantic Ocean by the extreme conditions, but survived in the Indo-West Pacific and were not able to re-establish themselves in the Atlantic Ocean, since there was not a land barrier between the two oceans.

What is the future of reef fishes in the West Indies? Some areas such as the North coast of Jamaica are severely overfished and large species, such as the groupers, are not abundant. The most common reef fishing gear used in the West Indies is the Antillean fish trap, a very efficient and effective device capable of overfishing an area when used in a high density and for a prolonged period of time.

Man influences reef fishes beyond simply over-exploiting them as food. Pollution of reef areas from dredging, sewage disposal, and industrial waste poses problems for the multitude of organisms dwelling on reefs. The health of the coral reefs themselves is critical to the continued survival and abundance of fishes in the West Indies.

☆

THE INTERNATIONAL OCEANOGRAPHIC FOUNDATION

JEAN BRADFISCH

Associate Editor "Sea Frontiers"

For those who love the sea, there are always mysteries to be unfolded, questions to be asked, and problems to be solved concerning it. Although the task is enormous, research is resolving some of the mysteries that have puzzled men for centuries. Questions about the myriad forms of life within it, about the dynamic forces that control its many moods, and about man's ability to cope with, utilize and conserve its life-giving and life-enhancing properties are beginning to be answered.

To afford laymen an opportunity to learn, in an interesting way, basic facts about the sea, as well as latest discoveries by ocean scientists, the International Oceanographic Foundation (IOF) was founded in 1953 by Dr. F. G. Walton Smith. Located at Virginia Key, Miami, Florida, on the western edge of the Gulf Stream, the IOF has become the leading disseminator of oceanographic information for the public.

As part of its educational program over the years, the IOF has encouraged scientific exploration and basic research, maintained a scholarship fund, and sponsored a Nordic-American exchange program for marine biologists. To share the excitement of this rapidly growing field with its 70,000 members in over 90 countries, the IOF publishes *Sea Frontiers* and *Sea Secrets.*

The colorful and definitive *Sea Frontiers* has become one of the foremost popular publications about the oceans. Articles, many by scientific authorities, are well illustrated and written in easy-to-read language. Since our frontiers into the sea are being constantly pushed back as technology advances, each issue brings new information and new photographs of places, things, animals and phenomena heretofore unknown. The world's leading underwater photographers focus their lenses on the strange, the curious, and the beautiful of the marine world.

Sea Secrets is an oceanographic newsletter. Material used in it is derived from questions submitted by members. The most interesting questions of general interest and their answers are published. The special, illustrated section, "Sea Winds," keeps readers informed about IOF activities and briefly reports on current developments in oceanographic research throughout the world.

The IOF has now embarked on the most ambitious program in its history. Plans are being developed for Planet Ocean, a two-acre building that will house exhibits about the ocean—exhibits wherein the visitor will not only learn about the ocean but can experience it. Visitors will not be just viewers, but participants as well.

Planet Ocean will also serve as the permanent headquarters of the IOF and the center for an expanded educational program. This program will feature learning materials for everyone from preschoolers to senior citizens. Special SEAKITS of curriculum enrichment materials will serve all levels of schools. The basic curricula will also be assembled into exhibitions that can be used cooperatively by museums throughout the nation.

Water is our most valuable resource. Without it, life as we know it could not exist. Although science provides information and the IOF disseminates it, it is citizens who must make decisions. *Sea Frontiers* reaches over 70,000 people each issue, Planet Ocean will reach 300,000 to 400,000 each year, and SEAKITS will reach millions (especially those who will make the decisions of the future).

While at Planet Ocean, visitors will take a simulated voyage through the solar system, which will demonstrate that only in a very narrow band, between Venus and Mars, a mere 1/350 of our solar system, it is possible for Earth—the blue planet, the life-giving, ocean planet—to exist. Only informed citizens will be able to plan, protect, and use it wisely.

For information write: The International Oceanographic Foundation, Rickenbacker Causeway, Virginia Key, Miami, Florida 33149.

THE UNIVERSITY OF MIAMI'S SCHOOL OF MARINE AND ATMOSPHERIC SCIENCE

The Rosenstiel School of Marine and Atmospheric Science of the University of Miami provides a laboratory for the study of dynamics of physical and biological aspects of ocean systems, on a seven-acre waterfront campus located only a few miles from downtown Miami on Virginia Key. It is one of the world's largest marine educational and research establishments, and offers graduate degrees in the marine sciences.

The training of graduate students is greatly enhanced by the opportunity to engage in actual research programs being carried out in various divisions of the School. Research covers a wide spectrum of the marine sciences, including such subjects as the oceanic circulation of the central Atlantic, life histories of fishes and invertebrates, taxonomy and population dynamics, marine ecology, fisheries investigations in Florida and the West Indies, fish farming utilizing the effluent heat from a nuclear power plant, and studies of the propagation of underwater sound.

The School maintains a small fleet of oceanographic research vessels. Largest of the fleet, the 208-foot R/V James M. Gillis, has accommodations for nineteen scientists and twenty-two crew men. For research and training in ecology the University operates two field stations. One is at Pigeon Key, 100 miles south of Miami, and the other is in the Everglades National Park.

Seminars by leading oceanographers from other institutions are offered on a regular basis at the School. In addition, special conferences are held periodically so that researchers from other institutes, local businessmen and community leaders may have an opportunity for communication and idea exchange in the marine sciences.

The Gulf and Caribbean Fisheries Institute, founded by the School in 1947, brings together each year scientists, government officials and representatives of industry to discuss problems and developments connected with the fishing industries. The GCFI has attracted world-wide interest.

THE ROLE OF THE MIAMI SEAQUARIUM IN SCIENTIFIC RESEARCH AND EDUCATION

WARREN ZEILLER

Managing Curator, Miami Seaquarium

The Miami Seaquarium appeals to people of all ages and all degrees of education, stimulating in them an interest in the world beneath the sea. The dramatic killer whales and the flipping dolphins that live here are known throughout the world. Less known are the contributions that have been made to science and to education.

"The unique facilities of the Seaquarium entail a vital responsibility upon us to cooperate with the scientists and students, to provide specimens and facilities for research, to work with scientific institutions and schools and to provide adequate answers to the questions that we have stimulated," says Burton Clark, general manager of the Seaquarium.

This great tropical marine aquarium is an integral part of the unique Virginia Key marine science complex, which includes the University of Miami's Rosenstiel School of Marine and Atmospheric Science, the National Marine Fisheries Service and the Atlantic Oceanographic and Meteorological Laboratory. The Seaquarium has worked closely with scientists and students from all these institutions.

The list of cooperative research programs conducted here is long: study of shark behavior, taxonomy of fishes, bioacoustics, visual acuity of the bottle-nose dolphin, physiology and behavior of manatee, cancer research, blood studies, pompano-rearing experiments, and a host of other projects.

The Dade County Laboratory Research program was created by the Dade County Board of Education in 1966, and since that date the Seaquarium has cooperated fully with the project. It is designed to enable outstanding high school students who are majoring in sciences to conduct independent research, under supervision, as part of their high school curricula. Since the beginning of the project, students selected annually have worked on biological, biochemical and behavioral problems under my supervision Many have obtained college scholarships as a result of their Seaquarium research projects.

Thousands of Dade County public school classes visit the Seaquarium annually at low student rates, as a class project, and for many the staff prepares special programs.

We have worked with the Florida Department of Natural Resources for several years on a project to rear hatchling green sea turtles to restock Florida waters with this endangered marine turtle. Staff members have worked with the Department in formulating standards for the capture, maintenance and shipping of marine mammals from the state's coastal waters, standards that have served as a model for federal legislation. The Seaquarium has also cooperated with the U. S. Navy's Man-in-the-Sea project.

Identification photographs in color of more than 450 species of marine fishes and invertebrates are made available for reference to scientists and qualified students. Thousands of specimens of several hundred species of sea creatures are preserved at the Seaquarium and given to scientists and qualified students working on specific projects.

Certainly a major contribution has been that millions of people from all over the world have had the opportunity to learn of the fascinating beauty and diversity of the 20,000 creatures of the sea that live here. The questions we have stimulated, the interest and enthusiasm we have evoked, have contributed immeasurably to man's current drive to explore and understand the world beneath the sea.

THE BERMUDA AQUARIUM AND MUSEUM

JAMES BURNETT-HERKES
Curator, Bermuda Aquarium

Founded in 1926 and opened to the public two years later, the Bermuda Aquarium is one of the oldest in existence. The site chosen by the founder, L. L. Mowbray, is on the edge of a tideway boiling alternately in and out of Harrington Sound, a two square mile body of water. Surface water is pumped from this tideway into storage tanks on the roof then gravity fed to the exhibit tanks. This is all done without filtration which means that on stormy days (hurricanes) the water in the tanks becomes a bit cloudy and, even more important, natural growth of countless marine organisms develops in the twenty-seven tanks. The observant visitor is able to see the seasonal succession of marine algae, dominated in the winter and early spring by filamentous greens like *Enteromorpha* being replaced in summers by lush growths of *Dictyota* and *Caulerpa*. The sponges also show seasonal activities and "blooms" of blue-bleeder and chicken liver sponges are noticeable during the spring and summer. Little anemones *(Aiptasia)* spring up in the summer like a garden of flowers and feed on the plankton that continually flows past their waiting tentacles. Open systems with a natural influx of organisms are not without their problems, however, and the undesirable things like sea urchins *(Lytechinus)* that feed on the algae have to be continuously weeded out. Parasites of fishes are often introduced and since it is impossible to treat fish without harming the natural growth in the tanks, sick ones are released and replaced by others.

Usually from fifty to seventy species of Bermuda reef fish are on display and most live and grow in the tanks for a long time — one for thirty-three years before it leapt out of its tank onto the floor in the dead of night and so terminated its residency. Interactions between the many fish species in the 1,000 and 2,000 gallon tanks can be seen daily and the gaily colored damselfishes often prove the most interesting with their elaborate territorial, breeding and nesting displays. It is also common to see the intricate sequences that accompany parasite picking. The primary parasite cleaners in the tanks are the wrasses, in particular the blue-head wrasse *Thalassoma bifasciatum,* and the little semi-transparent anemone shrimp *Periclimenes.* Fish, too, show seasonal activity patterns, and on cold winter days the wrasses will burrow into the sand while many of the parrotfishes spin mucous cocoons beneath an overhang of the reef. Large pieces of reef corals are periodically placed in the tanks to duplicate the beauty found on Bermuda's reefs. The reefs are dominated by growths of beautiful gorgonian corals, the seafans and sea rods, and by massive stony corals like *Diploria* and *Montastrea.*

Active pelagic fishes like barracuda, jacks and sharks can be seen in the large (40,000 gallon) reef tank in company with relatively sluggish demersal groupers, snappers, hogfish, and moray eels. Small reef fishes, like grunts, porgy, damselfish, butterfly fish and silversides are periodically added to this system and natural predation ensues. The groupers are most active during the day, while the sharks and morays do most of their hunting at night, when, luckily, visitors are not normally present.

Scientists and students from the Bermuda Biological Station often conduct research at the Aquarium on topics like behavior, diurnal activity patterns or the physiology of sleep. Staff at the Aquarium have been the agency for research and development of Bermuda's fishing industry and are currently involved in projects like the study of oceanic whales (in particular the beaked whales, *Xiphius)*, marine turtles, and in cataloging Bermuda's fishes. The Island's Natural History Museum was added to the Aquarium in 1941 for displaying subjects of Bermudiana that could not be shown live in the tanks or in the small zoo located on the grounds. Since this was, for many years, Bermuda's only Museum, it has diversified and now contains, apart from a full spectrum of natural history displays, exhibits on marine archaeology including the beautiful and priceless 16th century treasure recovered by famed Bermudian diver Teddy Tucker.

With virtually all of the space at the Aquarium complex utilized for exhibits, plans for the future center on education and research and development of the fishing industry, although the present exhibits are continually being improved, enlarged and brought up to date.

THE HYDRO-LAB

ROBERT I. WICKLUND

Project Manager, Hydro-Lab

In February 1971 a 200-foot cargo ship with a large crane on board eased near a bright orange marker buoy 1.2 miles southeast of Bell Channell inlet off Lucaya, Grand Bahama Island. The crane slowly lowered a strange yellow cylindrical object to the ocean floor, fifty feet below. The Hydro-Lab undersea laboratory was launched.

Since that day, the Hydro-Lab Underwater Research Program has progressed far beyond its original purpose of providing an educational tool to familiarize the public with the possibilities of undersea living and working and to prove to the scientific community that undersea living does not have to be the expensive proposition it has always been in the past in necessarily short-lived programs.

In two-and-a-half years Hydro-Lab has housed more than 150 scientists and technicians, most of whom remained underwater for seven days at a time. This number represents three times as many saturated (i.e., staying underwater too long to surface without a lengthy decompression) divers as any other undersea habitat in the world can claim. At this time it is the only operational long-range habitat program anywhere in the world.

Hydro-Lab is an eight-foot high by fourteen-foot long steel cylinder on legs which, while far from containing the comforts of home, has fulfilled a dream for many dedicated scientists in diverse fields.

This tiny room fifty feet below the sea surface has enabled intensive study of the marine environment and man in this environment without the time limitations divers have had to accept in the past. A broad spectrum of research has been carried out in Hydro-Lab, including studies of the physiology of saturated divers, plankton distribution, reef chemistry, fish behavior and distribution, hydrographic data collection, psychology, pollution and cosmic rays. Several training programs have also been initiated.

The seas of the Bahamas offer an ideal location for a habitat program. Clear waters allow excursions away from the habitat and warm temperatures increase a man's total work output because he is able to remain in the water for long periods.

Hydro-Lab was built by Perry Oceanographics in Riviera Beach, Florida, and the research program was originally funded by John Perry, jr., president of Perry Oceanographics, and Wallace Groves, Freeport developer. Almost total financial support now comes from the National Oceanic and Atmospheric Administration of the U.S. Department of Commerce.

Hydro-Lab has provided convincing proof that important research can be carried out in a low-cost habitat operation. Its success has spoken for itself. The emphasis now is on advanced research and improved methods. The dream for the future is to have many such habitats operating all over the world, modified to every type of marine environment.

(Photograph of Hydro-Lab on page 45)

THE ISLAND RESOURCES FOUNDATION

From out of the depths of the seas of the West Indies there rises a remarkable archipelago of islands, and, like misplaced icebergs in a tropical sea, only their tips are normally visible to mortal man. Small, remote, fragile, often unique, sometimes mysterious, always compelling, the oceanic islands of the West Indies represent for mankind a precious natural and psychological resource. For centuries man has used these tiny outposts as waystations on voyages between continents, as pawns in warfare and politics, and as exploitable possessions. More recently, he has come to see them as a place to retreat from the harassments of modern urbanized life. Now, a virtual tidal wave of people is spilling out of the continental melting pot of the world, drawn by idealized concepts of tranquil green islands in the sun. This flood-like migration is generating far reaching, occasionally promising and potentially disastrous effects upon all aspects of island life, for resident and visitor alike.

Throughout the world, insular environments, with limited land areas, are experiencing the pressures of growth and development technologies which have rarely been equaled on larger land masses. Attracted by the undeveloped, pristine nature of islands and by their recreational qualities, more people and industries from larger continental areas are traveling to the shores of previously isolated insular environments, threatening to alter the very qualities—the sense of romance, mystery, the isolation, smallness—which make islands unique.

In effect, islands are an endangered species, and the Island Resources Foundation is committed to the premise that prompt, coordinated, creative action is required to enhance the chances for survival of island ecosystems and their surrounding undersea habitats and their respective inhabitants.

The Foundation was established in 1970 as a non-profit, independent research center for the study of island systems, dedicated to improved resources management, comprehensive planning, and the conservation of the cultural, physical and natural resources of islands.

The Island Resources Foundation is the only research/educational center in the world founded for the express purpose of working with small island communities. The focus is to assist islands in meeting those requirements for rational growth and environmental quality needed in an often highly irrational world of development pressures and conflicts which neither comprehends nor appreciates the necessary defensive strategies, programs, and plans required to insure indigenous insular environmental integrity and local cultural values.

With administrative headquarters in the U.S. Virgin Islands, the Foundation carries out:

(1) *Research Programs* in tourism, mariculture, reef ecology, marine pollution, island history and culture, and environmental policy and resource management.

(2) *Technical Assistance Programs* in environmental management, oil spill contingency planning, national park development, historic site restoration, environmental education, development planning, pollution control, museum development, and conservation.

Further information is available by writing Dr. Edward L. Towle, president, Island Resources Foundation, Post Office Box 4187, St. Thomas, U.S. Virgin Islands 00801.

THE INTERNATIONAL UNDERWATER EXPLORERS SOCIETY (UNEXSO)

ROBERT S. FARRELLY

Located on Grand Bahama Island in the Bahamas, the International Underwater Explorers Society is perhaps the finest example of a resort diving facility in the world today. Its reputation as a professional diving center is well known in diving circles and it has been a primary contributor to important educational programs present in resort diving today.

Probably the best known and highly successful program has been the three-hour introductory SCUBA (Self Contained Underwater Breathing Apparatus) Course in which over 15,000 students have been trained.

Perhaps the greatest misconception present in diving is that the participant must be a fine, physical specimen, with great swimming ability and strength. UNEXSO has proven that this is not at all the case. In its three-hour course it has successfully graduated, for instance, complete non-swimmers, amputees, paraplegics, and one blind person. The question of age is always brought up as well. The youngest person taught at UNEXSO was 5½ years old, and a grandmother of 75 years young completed the course and visited our reefs. At the Society there is a saying: "If you can breathe through your mouth, you can dive", and basically this is true. The major obstacle to overcome is not age or physical ability, but being mentally able to handle this new and mysterious adventure. Diving is safe, provided human limitations are realized and practiced. Diving certainly is much safer than skiing down a mountain covered with snow.

The uniqueness of UNEXSO not only comes from its highly trained instructors and guides, and its educational programs, but, as well, from its physical plant which boasts an 18′ training tank, shallow activity pool, locker rooms with saunas, Cousteau museum, library, lounge, underwater experience room, complete photo laboratory and specially equipped diving boats. It is also the headquarters for the Perry Hydro-lab which is presently the most used underwater habitat (living quarters and laboratory) for scientific study in the world to date. Saturation dives in the Hydro-lab allow scientists to live and work on the bottom (50 feet) for seven days at a time without returning to the surface. Experiments being performed here, as well as the knowledge being gained on man living under the sea, may well prove to be important guides for man's future beneath the ocean.

Recreational diving is rapidly becoming one of the world's most popular sports and may, within a decade, rival today's leading sports. It is for this purpose UNEXSO prepares itself through its educational programs and constant attempts to re-evaluate and improve itself. Because the only way to overcome the fear held by man for the sea is through education: UNEXSO, in the past three years, has programmed and put into practical application, basic courses in Marine Identification, Decompression Diving, Underwater Photography, Certification Programs in SCUBA and an advanced Course in SCUBA. All of these courses can be taken during a planned week's vacation (with the exception of the advanced course which is fourteen days), and there is still time to do other things as well. These courses are meant to be fun and educational, not long and tiring. UNEXSO also sponsors its annual workshop in diving medicine for physicians. This course, headed by Dr. Ed Tucker, brings in such notables as Dr. George Bond and Dr. Hans Kylstra for special sessions, and concentrates on making the physician more aware of medical problems in diving which, with the fantastic growth of the sport, is becoming a necessity. The Society, as well, brings in well-known photographers, four times a year, to teach in its underwater photography workshop.

Special summer programs in diving and marine biology are taught to groups of students so as to further their interest in our oceans. Special safari trips are run on a monthly basis throughout the Caribbean also. One of these takes students to Yucatan for a Marine Archeology Expedition, to study several old wrecks as well as the Mayan ruins in the area.

The Society, having been responsible for a good deal of support in the scientific field, also has assisted the television and movie world. Many sequences of television's "Primus" were shot with UNEXSO's support, as were various series of the "Wild Kingdom" and "The American Adventure."

Of course, all of these things could not be possible if it were not for the excellent coral reefs and clear water located off Grand Bahama Island. Very few places offer the excellent variety of reefs and reef life that can be found here. The proximity of UNEXSO to the three major reefs makes it possible to run three and four guided diving trips each day. These three reefs consist of shallow reef (6′ to 15′) alive with every kind of coral and subtropical fish; the medium reef (30′ to 50′) having large growths of coral shaped like giant mushrooms growing out of a lovely, white sand bottom; the deep reef (50′ to 85′) dominated by massive solid growths of plate corals and topped by gorgonian corals. The ledge drops off at 125′ and plunges to 2000′. It is here that the Society does its decompression diving, going to 250′, on some occasions. Here on the ledge, beautiful growths of coral, colourful and large sponges, and the mysterious, black coral abound.

It is an exciting and magnificent world on these reefs and an experience one should not miss. It is a simple matter to learn diving as long as professional instructors are available to teach. The International Underwater Explorers Society is specifically set up for you, whether a beginning diver or an experienced one. Its goal is to provide for you, as an interested individual, the opportunity to become more totally involved in understanding our world beneath the Seas.

☆

THE UNDERWATER HUNT

CAPTAIN WILLIAM B. GRAY

Director of Collections and Exhibits, The Miami Seaquarium

The collector who dives for rare and exotic little reef fishes found in the coral canyons around South Florida, the Bahamas, Bermuda and in the Caribbean resembles both a golfer and an underwater butterfly hunter. Like a golfer, he carries a variety of tools—nets of all sizes and types, slurp guns and crow bars. He tows his golf bag through the water—the bag being a wire net supported by an inner tube. He snorkles along in clear water, always on the lookout for the rare and beautiful little fish.

When he spots a prize, he grabs the right net, and dives, going after that one particular fish. Collecting the little jewel-like reef fish is not a matter of random scooping. The collector cannot dip out a pailful of fish, anymore than the butterfly collector can casually wave his net through the air and catch butterflies. It is a matter of pursuit of one individual specimen, following the fish around coral heads and through crevices until you net it or lose it.

Corals can only live in unpolluted sea water where fast flowing water supplies the sustenance they require. The polyps that live in and build the coral reefs cannot go out and search for food. They live on algae and microscopic animal life and require large quantities of calcium to build their reefs. The reefs are inhabited by myriads of colorful fishes which congregate among the coral branches where they can seek refuge from their enemies. Many species live among the coral, and all are endowed with speed and cunning.

Since I became director of collections and exhibits at the Miami Seaquarium when it opened in 1955, we have gone on many different kinds of sea-hunts to bring 'em back alive from the ocean. We have collected two-inch sea horses and one-ton sea-cows. We net porpoises with a custom-made net that is a half mile long and twenty-two feet deep, with a webbing of twelve-inch heavy nylon mesh. We trawl through turtle grass for sea horses. We set out fish traps, for which we have permits from the Florida Department of Natural Resources.

But the most fascinating form of hunting is diving in the coral canyons, pursuing the specimens we want in their own element and netting them with hand nets. In shallow waters, we use snorkels, in deeper waters SCUBA equipment. But either way it is not a simple matter of just going out and scooping up many exquisite varieties and it requires a lot of patience and ingenuity to be successful. We need many specific reef fish for the Seaquarium's colorful corridor tanks—different varieties of butterflyfish, the rock beauties, the queen and French angels, jewel fish, squirrel fish.

Among the rare and exotic specimens that were first exhibited at the Seaquarium were the pigmy angelfish. We dove them out of sixty-five feet of water on the other side of the Gulfstream.

Among the small fish that we especially prize are the little jawfish. These pearl-colored creatures are about three inches long, and they are fascinating to watch because they dig tunnels. On pebbly bottoms they excavate long tunnels by bringing pebbles out of holes in their mouths, and spitting them out to form a parapet around their holes. They back into these tunnels tail-first when they are threatened. Our visitors at the Seaquarium enjoy watching these energetic little cave-builders. They are found in water that is twenty feet deep or deeper, and are concentrated in certain areas where the bottom is right.

Another prize is the little *Octopus vulgaris*, found on both deep and shallow reefs. An octopus will retreat rapidly into a hole if you pursue it, and it may be difficult to get it out of the hole. The secret of getting it out is to dump salt on the octopus. You take down a bottle of dry salt and pour it in the hole and the octopus will come out flying.

Once we were collecting in a bay in the Bahamas that was littered with trash. We saw a nice octopus retreat into a hole, and I went topside to get the salt, leaving my assistant to watch the hole.

In a few seconds my assistant surfaced, and shouted, "Capt'n, that octopus came after me! He came out of his hole, grabbed a broken bottle by the neck, and came after me!" An eight-armed barroom brawler wielding broken bottles could be really dangerous.

What of the dangers to the collector? I'd say that the scorpion fish were the most dangerous, because they are so camouflaged that it would be easy to brush one or step on one. The scorpion fish, which reaches about a foot in length, is difficult to recognize as a fish unless it moves. When lying still in its natural habitat it resembles a mossy rock or sponge. Near the points of the numerous needle-like spines in its dorsal fin are small holes from which the fish injects a deadly poison when the spine pierces the flesh of a human or predator attacking it. A slight prick on the hand is most excruciating and there are records of humans who have died after getting a large dose of the venom.

What about sharks? We don't worry about them. I have had fifty-five stitches taken in my leg due to a shark bite, but that shark was on land. I was helping take it from the collecting vessel to the shark channel when it twitched its head suddenly and carved a long slice in my leg. We consider sharks dangerous only in murky waters where they feel half-hidden. But in clear waters, where we collect and where the shark can see the whole person of the diver, they do not attack. At least they have never attacked me except on land, and I have been diving in the Atlantic, Caribbean and Pacific waters for more than fifty years.

Experienced divers don't consider barracudas dangerous in clear waters. A barracuda is curious; he will follow you around with his mouth half open as though he had adenoids, and gets close enough so that you can count all those sharp teeth. But he seems to be simply curious, and will go away after he has looked you over.

We are always careful where we put our hands on a reef, because moray eels lie hidden in holes, and can dart out and give a nasty bite if they feel threatened. The moray eel does not inject a venom, but those puncture wounds are very prone to become infected. The sea urchins with long, black, brittle spines can be very painful if you touch one of those spines and it breaks off in your foot or hand. Rubbing against fire coral can burn, and when we do chance to contact fire coral we immediately wash the surface of our skin with ammonia or gasoline.

Live coral reefs around the continental United States are found only off the coast of South Florida. They begin about three miles off shore from Miami and continue in a broken chain along the edge of the life-giving Gulf Stream on south beyond Key West. Florida has set aside a vast area of the offshore coral reef to conserve and protect sea life. The Key Largo Coral Reef Preserve, also known as Pennekamp Park, begins about sixty miles south of Miami and extends south to Molasses Reef Light and from there four miles shoreward to a beacon off Rock Harbor. Boating, angling and SCUBA diving are permitted, but spear-fishing is banned and it is unlawful to damage or remove corals, sea fans, sea plumes and crustacea. This marvelous underwater dreamland is a favorite part of the world for undersea photographers.

The Seaquarium does not, of course, collect in the undersea park, but there are miles of fine reefs south toward Key West where we can go hunting for the little exotic reef fish. We also collect in the crystal clear waters of the Bahamas. From Bimini south all the way to Riding Rock there are magnificent coral reefs around the Bahama Islands. Some of the most beautiful are in the Tongue of the Ocean east of Andros Island.

Years before the Seaquarium was created I collected all the way across the Caribbean from Jamaica to Mexico and Central America with the George Vanderbilt expedition. We were collecting fish specimens for the Academy of Natural Sciences of Philadelphia. There are grand reefs around Jamaica, Swan Island, and the Cayman Islands, and the Cayman reefs are also rich in wrecks. The Pedro Cays and the Serranilla Bank south and southwest of Jamaica are magnificent. From Cozumel, a Mexican Island in the Gulf of Mexico, south to Panama I have collected on extensive reefs. We netted many rare specimens off Honduras and Nicaragua. However, when we reached the reefs off the San Blas Indian territory off the coast of Panama we found gorgeous underwater scenery, but fish were scarce as hen's teeth. Not every reef brings you rewarding spoil.

ACKNOWLEDGMENTS

In preparing this volume I had the valuable assistance of several scientists and experts who contributed in their own field and checked my part in this book. The leading help I received from **Robert C. Work.** With his great all around knowledge of underwater life, he checked and often corrected my contributions to the last details. He contributed two important chapters to this book about the sponges and the mollusks. The exact identifications of subjects in the pictures were made by him, **Frederick M. Bayer, Patrick L. Colin** and **Dennis M. Opresko.** Patrick Colin wrote the story about reef fishes and Dennis Opresko the two chapters about corals and about gorgonians.

My appreciation goes also to **F. G. Walton Smith,** president of the International Oceanographic Foundation, for contributing the photographs to the coral story and material, especially two color pictures from "Sea Frontiers." Great assistance I was given also by **Burton Clark,** General Manager, and **Warren Zeiller,** Curator of the Miami Seaquarium.

For composing the picture part I had help by **Dave Woodward** with his thousands of underwater transparencies and **Jim Latourrette,** chief photographer of the Miami Seaquarium. **Bernd Mock** and **Armando Jenik** offered their fine collection of excellent reef pictures with interesting information. The **International Underwater Explorers Society** supplied the pictures of the underwater vessels.

In developing this big volume from my small handbook "Beneath the Seas of the West Indies," I had the editorial assistance of my experienced collaborator **Jane Wood Reno** and of **Lisa Drew,** editor of Doubleday & Company.

Artist **Bill Hays** designed the cover; my wife **Ilse** composed the final layout of this book; **Fer-Crom** made the excellent color picture separations; **Graphic Composition Services** set the type, and **Industria Grafica Domingo,** with its expert **Antonio Peña** did the printing.

My heartfelt appreciation goes to everybody who assisted in the preparation of this book, which I hope will be attractive and will bring more interest and knowledge to many people about life and beauty beneath the surface of the tropical and subtropical Western Atlantic.

Hans W. Hannau